IoT and Big Data Analytics for Smart Cities

The book *IoT and Big Data Analytics (IoT-BDA) for Smart Cities – A Global Perspective,* emphasizes the challenges, architectural models, and intelligent frameworks with smart decision-making systems using Big Data and IoT with case studies. The book illustrates the benefits of Big Data and IoT methods in framing smart systems for smart applications. The text is a coordinated amalgamation of research contributions and industrial applications in the field of smart cities.

Features:

- Provides the necessity of convergence of Big Data Analytics and IoT techniques in smart city application.
- Challenges and Roles of IoT and Big Data in Smart City applications.
- Provides Big Data-IoT intelligent smart systems in a global perspective.
- Provides a predictive framework that can handle the traffic on abnormal days, such as weekends and festival holidays.
- Gives various solutions and ideas for smart traffic development in smart cities.
- Gives a brief idea of the available algorithms/techniques of Big Data and IoT and guides in developing a solution for smart city applications.

This book is primarily aimed at IT professionals. Undergraduates, graduates, and researchers in the area of computer science and information technology will also find this book useful.

IoT and Big Data Analytics for Smart Cities

A Global Perspective

Edited by
Sathiyaraj Rajendran
Munish Sabharwal
Gheorghita Ghinea
Rajesh Kumar Dhanaraj
Balamurugan Balusamy

CRC Press
Taylor & Francis Group
Boca Raton London New York

CRC Press is an imprint of the
Taylor & Francis Group, an **informa** business

A CHAPMAN & HALL BOOK

First edition published 2023
by CRC Press
6000 Broken Sound Parkway NW, Suite 300, Boca Raton, FL 33487-2742

and by CRC Press
4 Park Square, Milton Park, Abingdon, Oxon, OX14 4RN

CRC Press is an imprint of Taylor & Francis Group, LLC

Library of Congress Cataloging-in-Publication Data
Names: Rajendran, Sathiyaraj, editor.
Title: IoT and big data analytics for smart cities : a global perspective /
edited by Sathiyaraj Rajendran, Munish Sabharwal, Gheorghita Ghinea,
Rajesh Kumar Dhanaraj, Balamurugan Balusamy.
Description: First edition. | Boca Raton : Chapman & Hall/CRC Press, [2023]
| Includes bibliographical references and index. |
Identifiers: LCCN 2022025755 (print) | LCCN 2022025756 (ebook) | ISBN
9781032108551 (hbk) | ISBN 9781032380490 (pbk) | ISBN 9781003217404 (ebk)
Subjects: LCSH: Smart cities. | Internet of things.
Classification: LCC TD159.4 .I675 2023 (print) | LCC TD159.4 (ebook) |
DDC 307.760285--dc23/eng/20220906
LC record available at https://lccn.loc.gov/2022025755
LC ebook record available at https://lccn.loc.gov/2022025756

ISBN: 978-1-032-10855-1 (hbk)
ISBN: 978-1-032-38049-0 (pbk)
ISBN: 978-1-003-21740-4 (ebk)

DOI: 10.1201/9781003217404

Typeset in Palatino
by SPi Technologies India Pvt Ltd (Straive)

Contents

Preface..vii

Editors..ix

List of Contributors..xi

1. **Era of Computational Big Data Analytics and IoT Techniques in Smart City Applications**..1
   T. Parameswaran, Y. C. A. Padmanabha Reddy, V. Nagaveni, R. Sathiyaraj

2. **Challenges and Roles of IoT and Big Data Analytics Enabled Services in the establishment of Smart Cities**..23
   J. Jeya Bharathi, S. S. Aravinth, S. Madhusudhanan, U. Rahamathunnisa, N. Sudhakar Yadav

3. **Security and Privacy Challenges and Solutions in IoT Data Analytics**....................43
   Kumar Shalender and Rajesh Kumar Yadav

4. **IoT-BDA Architecture for Smart Cities** 57
   Garima Pandey, C. Ramesh Kumar, Mayank Kumar, Kashish Gupta, Shivangi Singh Jha, Jatin Jha

5. **Intelligent Framework for Smart Traffic Management System: Case Study** 75
   K. Aditya Shastry, H. A. Sanjay, M. Lakshmi

6. **IoT and Big Data Analytics-Based Intelligent Decision-Making Systems** 101
   N. Sudhakar Yadav, Sreenivasulu Gogula, Ganti Krishna Sharma, Ch. Mallikarjuna Rao, D. V. Lalita Parameswari

7. **Recent Advancement in Emergency Vehicle Communication System Using IOT** 121
   Ajay Sudhir Bale, Vinay Narayanaswamy, Varun Yogi Shanthakumar, Parinitha Balraj Shyla, Shivani Balakrishna, Varsha Shyagathur Nagaraja, Sahana Basatteppa Menasinakai, Manish Kumar Hitesh, Eshwar Esarapu

8. **Pandemic Management Using Internet of Things and Big Data – A Security and Privacy Perspective**..159
   K. S. Arvind, S. Vanitha, K. S. Suganya

9. **Sustainable Efficient Solutions for Smart Agriculture: Case Study** 175
   N. Sudhakar Yadav, Murali Krishna, I. Sapthami, Ch. Mallikarjuna Rao, D. V. Lalita Parameswari

Index..201

Preface

With advanced technologies, traditional human life has been migrated to a digital world. It has been observed that many of the advanced applications have been developed with the integration of advanced smart technologies like artificial intelligence, cognitive computing, machine learning, Big Data analytics, business intelligence, and Internet of Things (IoT). Among these technologies, Big Data Analytics with the integration IoT, offers extensive variety of applications and plays a vital role in assisting and managing smart city applications.

Nowadays, IoT devices are widely used in human's day-to-day life activities and in solving real-time applications. These devices generate huge amount of data and all these data have to be stored and processed using Big Data algorithms. IoT and Big Data converge together to provide energy efficient solutions for smarter real-world applications.

Internet of Things (IoT) has transfigured traditional human life to smart life system. The goal of IoT is to "connect the unconnected". The unpredictable growth in the number of devices connected to IoT and the rapid increase in the volume of data generated, this evidences the necessity of Data Analytics for various IoT applications. There are various applications that are yet to be solved and aim to offer a potential solution for Smart City. IoT plays a predominant role in the growth of smarter applications in various domains. Moreover, we require the support of Big Data analytics in analysing and processing huge datasets in a secure, safe and efficient manner. With the use of data analytics algorithms, we can develop applications for smarter cities like traffic prediction system, smart traffic management, smart parking system, smart healthcare, smart waste disposal system, smart environment, etc. These applications may not be possible without the support of IoT, which connects, coordinates, and communicates the devices namely "things". The convergence of IoT and Big Data is essential in providing energy-efficient solutions for smarter cities. This addresses the necessity of integrating IoT and Big Data in providing energy-efficient solutions for smart cities. This book investigates the recent technological advancements and achievements in the said area and also it aims to discover the recent trends in IoT and Big Data Analytics to push the smart city concepts to the next level.

The primary objective of this book is to offer the recent trends in the field of IoT integrated with Big Data Analytics (IoT-BDA), and to dig out various energy-efficient solutions for promoting smart city applications. This investigates the recent technological advancements and achievements in the said area and also it aims to discover the recent trends in IoT and Big Data Analytics to push the smart city concepts to the next level. This book will also provide the future research directions of IoT-BDA in terms of offering better solutions for smart cities.

The book initiates the discussion with the era of computational Big Data Analytics and IoT techniques in smart city applications, thereby emphasizing the challenges of IoT and BDA in the establishment of smart city applications. Privacy concerns are addressed with an advanced architectural model deploying IoT-BDA. Intelligent framework for smart traffic management system is particularized in detail with a case study.

The book also covers intelligent decision-making systems experimented with a case study. Emergency vehicle communication framework is provided with the advanced concepts of IoT. How these techniques assist in the pandemic management is also briefly discussed. Lastly, the book concludes with sustainable efficient solutions for smart agriculture. The research results were demonstrated with a real-time case study.

Editors

Dr. Sathiyaraj Rajendran is an assistant professor in the School of Engineering & Technology at the CMR University, Bangalore. He completed his PhD at Anna University, Chennai. His research interests lie in the area of Big Data Analytics, AI, and IoT. He has collaborated actively with researchers in several other disciplines of computer science, particularly traffic prediction systems and intelligent systems. He has authored more than 25 publications and filed 5 patents.

Dr. Munish Sabharwal is professor and dean, School of Computing Science & Engineering, Galgotias University, Greater Noida (UP), India. He is also an adjunct professor, Faculty of Applied Mathematics and IT-Samarkand State University, Samarkand, Uzbekistan. He earned a PhD (Computer Science S) and PhD (Management) and also has a PGDM (International Trade) from Symbiosis Institute of Management Studies, Pune; an M.Tech (Computer Science) from IASE University, Sardarshahr; and a B.E. (Computer Technology) from Nagpur University, Nagpur. He is also MCSD and MCP Certified. He has contributed over 21 years to teaching (CS and MIS), education management in leadership roles, research, as well as software development. He has more than ten years of research experience, has published more than 55+ research papers in conferences and journals indexed in SCI, ESCI, Scopus, etc. He is a board member as well as reviewer for a number of leading indexed research journals and has three books to his credit. He is guiding seven PhD Scholars and his current research interests include Data Sciences (AI & ML), Biometrics and E-Banking.

Dr. Gheorghita Ghinea is a professor of computing, Department of Computer Science Brunel University London. He holds a PhD in Computer Science from the University of Reading, United Kingdom. His research activities lie at the confluence of Computer Science, Media and Psychology and he is particularly interested in building semantically underpinned human-centred e-systems, particularly integrating human perceptual requirements. Currently, he is supervising a team of six PhD students. He has supervised a total of 23 PhD students to completion (as first supervisor), who have gone on to enjoy rewarding careers in academia or industry. He has published more than 30 articles and received 10+ research grants.

Dr. Rajesh Kumar Dhanaraj is an associate professor in the School of Computing Science and Engineering at Galgotias University, Greater Noida, India. He earned a PhD in Information and Communication Engineering from the Anna University Chennai, India. He has contributed 20+ books on various technologies and 35+ articles and papers in various refereed journals and international conferences and contributed chapters to the books. His research interests include Machine Learning, Cyber-Physical Systems and Wireless Sensor Networks. He is an Expert Advisory Panel Member of Texas Instruments Inc., USA.

Dr. Balamurugan Balusamy earned a PhD in Computer Science and Engineering from Vellore Institute of Technology (VIT) Vellore, Tamil Nadu. He also completed an ME in Computer Science and Engineering and a BE in Computer Science and Engineering from Anna University and Bharathidasan University respectively. Previously he was associated with Galgotias University, Delhi NCR, in multiple capacities, holding roles such as Director of International Affairs, Professor, and Associate Dean of Research, School of Computing Science and Engineering. He served up to the position of associate professor during his 12-year stint with VIT (Vellore) from 2005 to 2017. During this period, he spearheaded and coordinated many student club activities and technical and cultural fests for VIT. Currently, he is appointed to the advisory committee for several technology start-ups and forums.

Dr. Balusamy promotes student engagement in co-curricular and research activities and is keen to provide them an ecosystem for redefining their career goals. He designs and implements activities for developing and overseeing programs that enable students to realize their social and personal potential and integrate student life experiences at Shiv Nadar University, Delhi NCR. In his current role at the University, he acts as a liaison between the Office of Dean Academics and the Office of Students. He mentors students with the twin aims of fulfilling their ambitions and helping them realize their full potential.

List of Contributors

K. Aditya Shastry
Nitte Meenakshi Institute of Technology
Yelahanka, Bengaluru, Karnataka, India

S. S. Aravinth
Koneru Lakshmaiah Education Foundation
Vaddeswaram, Andhra Pradesh, India

K. S. Arvind
Jain University
Jakkasandra Post, Ramanagara District
Karnataka, India

Shivani Balakrishna
CMR University
Bengaluru, Karnataka, India

Ajay Sudhir Bale
New Horizon College of Engineering
Bengaluru, Karnataka, India

Parinitha Balraj Shyla
CMR University
Bengaluru, Karnataka, India

Eshwar Esarapu
CMR University
Bengaluru, Karnataka, India

Sreenivasulu Gogula
Dean, Research and Development at ACE
Engineering College
Hyderabad, Telangana, India

Kashish Gupta
Chief operating Officer, Doubtfree EdTech
Private Limited
Noida, India

J. Jeya Bharathi
Koneru Lakshmaiah Education Foundation
Vaddeswaram, Andhra Pradesh, India

Jatin Jha
PR Manager of Doubtfree EdTech Private
Limited
India

Murali Krishna
PBR Visvodaya Institute of Technology &
Science
Nellore, Andhra Pradesh, India

Mayank Kumar
CEO of Readycoder Private Limited
Founder & CEO of Doubtfree EdTech
Private Limited
Chief Operating Officer of Indo Biopearl
Healthcare & Research
Chief Operating Officer of Foodenia,
India

Manish Kumar Hitesh
CMR University
Bengaluru, Karnataka, India

Rajesh Kumar Yadav
Amity University
Uttar Pradesh, India

M. Lakshmi
Nitte Meenakshi Institute of Technology
Yelahanka, Bengaluru, Karnataka, India

D. V. Lalita Parameswari
G. Narayanamma Institute of Technology
and Science (For Women)
Hyderabad, Telangana, India

S. Madhusudhanan
Prathyusha Engineering College
Thiruvalluvar, Tamil Nadu, India

Sahana Basatteppa Menasinakai
CMR University
Bengaluru, Karnataka, India

V. Nagaveni
Acharya Institute of Technology
Bangalore, Karnataka, India

Vinay Narayanaswamy
CMR University
Bengaluru, Karnataka, India

Y. C. A. Padmanabha Reddy
B V Raju Institute of Technology
Telangana, India

Garima Pandey
Galgotias University, Greater Noida
Uttar Pradesh, India

T. Parameswaran
CMR University
Bangalore, Karnataka, India

U. Rahamathunnisa
Vellore Institute of Technology
Vellore, Tamil Nadu, India

C. Ramesh Kumar
Galgotias University, Greater Noida
Uttar Pradesh, India

Ch. Mallikarjuna Rao
Gokaraju Rangaraju Institute of
    Engineering and Technology
Hyderabad, Telangana, India

H. A. Sanjay
M S Ramaiah Institute of Technology
Bengaluru, Karnataka, India

I. Sapthami
Visvodaya Engineering College, Kavali
Nellore Dist, Andhra Pradesh, India

R. Sathiyaraj
CMR University
Bangalore, Karnataka, India

Kumar Shalender
Chitkara University
Punjab, India

Varun Yogi Shanthakumar
CMR University
Bengaluru, Karnataka, India

Ganti Krishna Sharma
ACE Engineering College
Hyderabad, Telangana, India

Varsha Shyagathur Nagaraja
CMR University
Bengaluru, Karnataka, India

Shivangi Singh Jha
Chief Business Development Officer,
    Doubtfree EdTech Private Limited
Chief Technology Officer, Foodenia
Noida, India

N. Sudhakar Yadav
VNR Vignana Jyothi Institute of
    Engineering and Technology
Hyderabad, Telangana, India

K. S. Suganya
Bannari Amman Institute of Technology
Sathyamangalam, Tamilnadu, India

S. Vanitha
PES University
Bangalore, Karnataka, India

1

Era of Computational Big Data Analytics and IoT Techniques in Smart City Applications

T. Parameswaran
CMR University, Bangalore, India

Y. C. A. Padmanabha Reddy
B V Raju Institute of Technology, Narsapur, India

V. Nagaveni
Acharya Institute of Technology, Bangalore, India

R. Sathiyaraj
CMR University, Bangalore, India

CONTENTS

1.1 Introduction 2
1.2 Work in Progress in the Background 3
    1.2.1 Big Data Analytics 6
    1.2.2 Internet of Things (IoT) 7
        1.2.2.1 The Smart House 9
        1.2.2.2 Smart Cities are Places Where People Live, Work and Play 9
        1.2.2.3 Retailing in the 21st Century 10
        1.2.2.4 The Smart Grid 10
        1.2.2.5 Healthcare 10
        1.2.2.6 Poultry and Agricultural Production 10
1.3 Challenges 10
    1.3.1 Adding Value to the Customer's Experience 11
    1.3.2 Analytical Challenges 11
1.4 Smart City Applications 12
    1.4.1 Challenges Faced by Smart City Applications 13
    1.4.2 The Necessity of Integrating Big Data and IoT 14
    1.4.3 To Use IoT with Big Data to Solve the Problems of Intelligent Cities 15
1.5 Case Study 16
    1.5.1 Analyse the Historical Context (Prescriptive) 18
1.6 Final Dashboard 20
1.7 Conclusion 20
References 21

DOI: 10.1201/9781003217404-1

1.1 Introduction

Things connected with the Internet are pushing people to the brink of insanity. The Internet of Things (IoT) is a medium that allows cooperation and communication among various things to take place via the use of the cyberspace. The development in current technologies is having a significant impact on people's expectations for daily conveniences in a positive manner. It encompasses the fields of medical services, mechanisation, transportation and crisis response to various crises that occur when humans become unable to make decisions for themselves. The traditional definition of the term "Internet" (a network of computers) will never be regarded again. The IoT is an accumulation of billions of smart devices in addition to the established frameworks, which may result in an increase in the size and scope of the IoT, providing another technique for dealing with situations and also problems. Following the completion of the administrative level phases, the vast majority of countries have moved ahead with their national systems in order to realise IoT [1].

Several IoT applications are now being developed, with smart cities being the most notable of them. To guarantee high-quality service delivery across the community, novel applications have been developed and promoted in metro metropolitan regions, despite the high levels of urbanisation and population increase seen there. When information and communication technology (ICT) was used in municipal service supply, the terms "telecity," "smart city," and "information city" were created to describe the phenomenon. Researchers and scientists are motivated by the rapid growth of IoT technologies in the sense of developing new application areas and new IoT services, and these new smart services should satisfy the requirements of people all over the globe to a great extent. Human needs will also be taken into account via the exchange and collection of data within IoT services in order to increase awareness of smart city ideas across the globe. As a result, the network should include actuating, networking, processing and sensing components. Monitoring, gathering, archiving and sharing open sensor data from IoT devices are also essential objectives to achieve in order to assist the creation and study of smart cities [2]. In Figure 1.1, you can see the general layout of a typical smart city architectural design. At the outset, the Quality of Life (QoL) enhancement of urban residents and workers is the most important objective of smart cities. Numerous studies and research have been conducted by experts in the field of smart city design and implementation in real-world settings.

The vast majority of the projects are geared at improving or incorporating one or more of the critical components of a smart city, such as parking management, trash disposal and

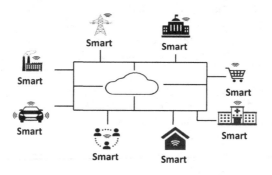

FIGURE 1.1
Conventional smart city architecture overview.

recycling, infrastructure management and so on. Nonetheless, a lack of integrity in the delivery of municipal services has a detrimental impact on the effectiveness of such services. The contemporary urbanised environment, as a result, exhorts us to meet the pressing needs for comprehensive smart city design that is adaptable and efficient [3].

Big data has been effectively used in a variety of industries, including public services, waste management, law enforcement and energy management. As a result, it has the potential to have an effect on sustainability in any industry. In a growing number of industries, researchers and policymakers are turning to big data, which is often produced by mobile phone users, social networking sites, different government and commercial websites, corporate software, everyday home appliances and other smart devices [4, 5].

This chapter gives an outline of how the IoT and big data analytics may assist in the development of smart cities. The primary emphasis will be on smart city problems that can be solved via the integration of those both technologies. The in-depth investigation of background work offers a comprehensive multi-disciplinary perspective of the problems encountered by the smart city communities, as well as the ways in which the data mining and knowledge management communities may contribute to solving those challenges. In addition, a comprehensive explanation of a case study involving the usage of big information and cyberspace for smart city construction is provided.

1.2 Work in Progress in the Background

Shafique and colleagues provided a comprehensive and in-depth analysis on forthcoming IoT-5G environment. Fifth Generation (5G) networks are critical enablers for the widespread IoT deployment. Examples of such technologies are aggregation of carrier, M-MIMO massive input multiple output, device-to-device communications, MIMO multiple input multiple output, CoMP – synchronised multipoint processing, CRAN – centralised radio access network, SD-WSN – software-defined wireless sensor networking, NFV – network function virtualisation, cognitive radios and CRNs –cognitive radio networks. They also looked at the new and the developments of 5G-IoT, which are being determined by advancements in reproduction intellect, machine learning and deep knowledge, as well as current 5G efforts, service quality needs in 5G and challenges related to standardisation [5]. Sudhakar et al. provided a comprehensive and in-depth analysis on forthcoming IoT environment. They designed the framework suitable for various applications ranging from tracking the status of autistic people to prediction of various diseases. Different sensors are integrated to the mobile to take the health parameters from the patient to monitor the data actively [6].

With the use of a mix of cooperative control and state prediction methods, Vara et al. explored scattered continuous-time fault assessment for numerous devices in IoT networks. For starters, a mode-based midway temperature matrix is created, in that an intermediate evaluator may be constructed to evaluate incorrect temperature data received from IoT. Further, the transition matrix and output temperatures of continuous-time Markov chains, as well as the necessary and sufficient criteria of stability for self-correct error of the temperature of IoT network, are taken into consideration. Furthermore, broken devices are interchanged with virtual devices for guaranteeing ongoing and strong observation of IoT network, thus avoiding fraudulent data gathering from taking place [7].

To better understand the key problems connected with the complex process of investigations based on IoT, Stoyanova et al. conducted a comprehensive research in which they addressed all challenges such as cloud security, privacy etc. In addition, this study offers an outline of the theoretical models used in the research field of digital forensics in the past and now, as well as future directions. The guidelines that seek for data extraction in a privacy-conserving way or else to protect the integrity of evidence by utilising decentralised solutions based on blockchain are given particular consideration in this section. In addition, the current Forensics-as-a-Service (FaaS) paradigm and several potential cross-cutting data reduction and intelligence methods related to forensics are discussed in this chapter. Last but certainly not least, many additional research directions and open challenges are discussed, with particular importance placed on the requirement of proactive Forensics preparedness initiatives and widely accepted metrics [8].

A review of IoT implications in the healthcare sector that are being practised lively and investigation were made by Lin et al. They presented three scenarios in which they reviewed the current direction and emphasis of healthcare applications of IoT:

1. Treatment of acute diseases. We will present three applications that will demonstrate how the IoT may improve acute care: essential sign monitoring, telemedicine, diagnosis based on IoT and management of communicable illnesses, among others.

2. Treatment of persistent diseases. The chapter focuses on telehealth monitoring, which is used to monitor the health of patients with chronic illnesses, particularly those suffering from heart failure, diabetes, or Alzheimer's disease.

3. Auto-management of one's own health. The chapter focuses on smart watches, which are the most widely used dedicated device for the management of self-health. It examines the most important features of smart watches in terms of self-health, namely tracking sleep and tracking exercise, and discusses how these functions can be combined.

Mirani et al. investigated the use of IoT in agriculture, as well as the difficulties that it presents. When you think of the IoT, you probably imagine embedding a framework into a device by adding computer software, sensors and actuators, and then connecting that item to the Internet so that data can be uploaded and analysed in the cloud. IoT devices, on the other hand, needless computing power, have a shorter battery life, support heterogeneity, are platform independent and use RFID, Wi-Fi and Bluetooth. As a result of this support, IoT devices became more suitable for remote sensing. It is used in the decision-management process to provide fast and prompt reaction. There are several agricultural applications for microcontroller technology, including IoT devices for more semantic monitoring, processing and uploading [9], which benefits the farm automation sector and crop products. Microcontroller technology has several agricultural applications, including IoT devices for further monitoring, processing and upload semantics [10, 11].

Kumar et al. looked at how big data analytics may be used in the healthcare industry. The healthcare business has had to contract with enormous capacities of information created from a number of foundations, all of which are well known for producing big amounts of heterogeneous data in large quantities. To deal with the massive amounts of data generated in the healthcare business, many big data analytics tools and methodologies have been industrialised. They discussed the impact of big data in healthcare, as well as the many Hadoop ecosystem tools for dealing with it [10, 11].

Gofrani et al. presented a study on current applications of big data in railway engineering. The study addressed three parts of railroad conveyance where BDA has remained used, namely processes, care and security. This study similarly evaluates and describes the degree of big data analytics, the kinds of big data replicas and the diversity of big data approaches. This research revealed that there are research gaps in BDA for railway transportation systems [11], which should be addressed in future BDA research.

For intelligent transportation systems, Zhu et al. examined the frameworks of big data that may be applied. They described the many data sources and collecting techniques, data analytics methodologies, as well as the categories of big data analytics application. They presented a number of case educations of big data analytics applications in intelligent transportation systems in their study [12], which included road circulation accident examination, circulation flow forecast, public transportation service preparation, individual portable way preparation, rail transport organisation and switch and advantage upkeep.

The creation of smart cities is intended to mitigate the difficulties that have arisen as a result of the ongoing growth of urbanisation and the growing density of population in cities. For addressing these difficulties, decision-makers and governments embark on initiatives of smart city that aim to promote sustainable economic development while also the quality of life improvement for both tourists and residents. SCDAP stands for the Smart City Data Analytics Panel, and it is a new big data analytics platform for smart cities developed by Osman and colleagues. According to their findings, they developed the SCDAP in order to answer the following research queries: What are the nature of big data analytics guidelines implemented in smart cities that have been published, and what are the vital principles of design that guide the big data analytics framework designs that have been developed to provide the purposes of smart cities? For addressing these concerns, they conducted a methodical evaluation of the publications on the frameworks of big data analytics in smart cities. New capabilities for the frameworks of big data analytics were introduced by the framework proposed [13], which are reflected in data model organising and accumulation.

To meet the current problems in information technology, Jeong and colleagues developed a variety of solutions, methods and frameworks that were previously unavailable. Such solutions include a wide range of upcoming track subjects, which includes blockchain, steganography, machine learning, smart systems, etc. Blockchain is only one example. In the following paragraphs, we will provide a succinct overview of each subject, including a discussion of the current problems and potential solutions. They also offered a variety of paradigms to topics that dealt with a variety of different types of study fields, such as the IoT and Smart Cities, among others [14].

When a big number of gadgets are introduced into the market in the near future, the potential of information leakage and privacy violation increases significantly. In order to avoid this, each device employs a privacy-preserving technique. In their paper, Gheisari et al. argue that all current methods suffer from three main flaws: (1) they use a single static privacy-preserving technique for the whole system; (2) they transmit the entire system's data at once; and (3) they are not context-sensitive. These have an unacceptably low level of privacy preservation. They begin by equipping smart cities based on IoT with the concept of Software Defined Networking to solve these issues. Afterwards, a very effective privacy-preserving mechanism was installed on top of it that controls the flow of data packets from divided IoT device data [15]. Further studies are being conducted into the use of the IoT and big data analytics in clever city requests.

1.2.1 Big Data Analytics

It was stated in early 2011 that big data will be the next cutting edge in terms of efficiency, novelty and competitiveness. In 2018, the Internet users raised over 7% in comparison to the previous year, reaching more than 3.7 billion people. In 2010, more than one zettabyte (ZB) data produced globally, by 2014, that figure had increased to seven ZB. Three Vs, namely volume, variety and velocity, were used to describe the developing big data features in 2001.

In 2011, the International Data Corporation (IDC) described big data with Four four Vs, namely volume, velocity variety and value. The term accuracy was proposed as the sixth feature of big data in 2012, and it is now considered essential. This figure shows the five most frequent characteristics of big data, which are shown in Figure 1.2. Volume is a term used in computing to indicate the size and breadth of a dataset. It alludes to the vast amount of data created every second. It's difficult to define a consistent threshold for big data volume (i.e., what makes a "large dataset") since the time of day and the type of data used might affect the meaning of "big dataset." Currently, datasets in the exabyte (EB) or terabyte (ZB) range are considered big data; nevertheless, smaller datasets still provide challenges. Variety refers to the many forms of data that may be discovered in a dataset, including structured, semi-structured and unstructured data. Structured data (such as data recorded in a relational database) is efficient and can be easily organised, whereas unstructured data (such as text and software content) is unorganised and difficult to analyse. Tags are used to separate data components in semi-structured data (e.g., NoSQL databases); however, the database administrator is responsible for maintaining this structure. Uncertainty can arise while converting between different data types (e.g., when converting from unstructured to structured data), expressing data that is a mix of data types and making changes to the dataset's underlying structure at run time. Because the data under observation comes from a variety of sources and is represented in a change of data types and symbols, efficient analysis on shapeless and semi-structured data may be challenging. The word "velocity" includes the speed (expressed in batch, near-real time, real time and streaming) with the focus on the fact that the speed with which the data is processed must equal the pace with which the data is created. The degree of truth reflects the data's quality

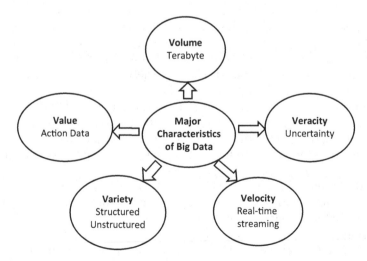

FIGURE 1.2
Five big data characteristics.

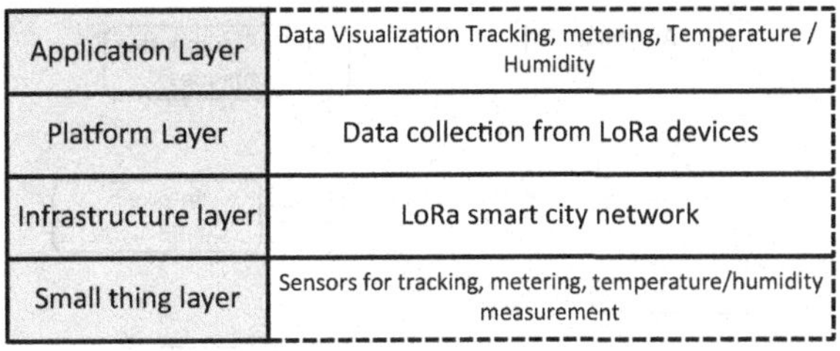

Application Layer	Data Visualization Tracking, metering, Temperature / Humidity
Platform Layer	Data collection from LoRa devices
Infrastructure layer	LoRa smart city network
Small thing layer	Sensors for tracking, metering, temperature/humidity measurement

FIGURE 1.3
Big data architecture.

(e.g., uncertain or imprecise data). Although worth signifies both the setting and utility of data for executive, the preceding V's are primarily focused on conveying the challenges connected with big data [16].

Figure 1.3 depicts the architecture of big data, which is divided into four layers: the layer of tiny things (STL), the layer of infrastructure (IL), the layer of platform (PL) and the layer of application (AL). The creation of the system architecture was guided by the concept of simplicity as its guiding principle. The design should not grow too complicated, since this may make deployment more difficult. The design makes use of a number of different software packages and also several hardwares such as sensors, devices used for data collection. The combination of all of the software tools was straightforward and never required the use of specialised knowledge. Additionally, the design should be readily expandable, and the installation of other modules should not be a problem.

For monitoring the parameters such as humidity, temperature, etc., IoST layer consists of sensors and LoRa devices. These devices produce huge amount of data. Every device produces a peculiar format of data containing information in Fahrenheit and Centigrade, also the details about the location of the device and its unique identification number, among other things. During operation, the gateway gathers all of the necessary information and transmits them to mainframe for processing further at the platform level.

The infrastructure layer contains a number of gateways for receiving data from the sensors placed and are utilised for various reasons. The LoRa network connects the devices via the usage of the Internet. Traditional data collection techniques are used to gather the first IoST data that is received at the platform layer of the stack. At this layer, a variety of problems are dealt with, including IoST data redundancy, noise reduction and small error correction, among others. Furthermore, pre-processing is carried out utilising the max–min normalisation method. The data from the IoST is visualised by the users at the application layer, using a variety of methods. The visualisation methods used must be simple to comprehend and effective in aiding decision-making. Simple tables, bar charts and graphs, as well as complicated yet relevant colouring schemes, are examples of presentation methods. The scale and units of measurement must be simple to comprehend.

1.2.2 Internet of Things (IoT)

The IoT's primary goal is to facilitate the sharing of real-time information among autonomous networked agents. Figure 1.4 depicts the IoT idea, with the system's inputs on the right side and its usage on the left side of the diagram.

It is necessary to install a sensor with sophisticated computational capabilities in a location that has access to the Internet. When this sensor is connected to the network, it will be able to interact with anything, at any time and from any location [17]. Data-collecting

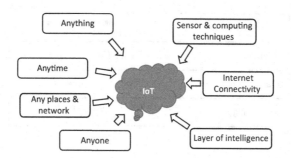

FIGURE 1.4
Concept of IoT.

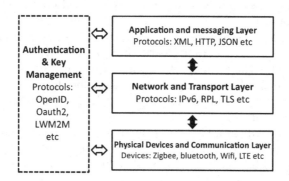

FIGURE 1.5
IoT architecture and corresponding protocols.

systems are used in the IoT architecture to discover and send data via a large number of such communication devices, simplifying the data collection process. Several communication options such as Wi-Fi, ZigBee, Bluetooth and GSM enable the connectivity of devices utilising various access networks such as Radio Frequency Identification (RFID), devices with wireless sensors and any smart item linked to the Internet through physical IP [18].

As shown in Figure 1.5, IoT protocols use several layers, each one for different applications, messaging, physical devices and routing/forwarding, also used in place of authentication and key management. IoT consists of Low-Power Wide-Area Network (LPWAN)-based protocols which are newly developed in addition to the Low-Rate Wireless Personal Area Networks (LR-WPANs) standards and protocols already available.

Two low-level layers such as the physical layer and the Medium Access Control (MAC) layer are available for LR-WPANs as mentioned in the standard IEEE802.15.4. The physical layer specification is concerned with the transmission of data over wireless channels that operate at various frequencies and data rates. The MAC layer standard is concerned with the mechanisms for channel access as well as channel synchronisation methods. Because the IEEE 802.15.4 standard has a small maximum transmission unit (MTU), an IPv6 over Low-Power Wireless Personal Area Network (6LoWPAN) adaptation layer must be added above the link layer to offer IP-based communication capabilities to sensor nodes. Each device in the IoT is identified by an IPv6 network address. The Routing Protocol for Low-Power and Lossy Networks (RPL), created by Cisco, supports six LoWPAN configurations. The RPL standard supports point-to-multipoint communication as well as communication between multiple points and a single point.

Because of the limited payload, the UDP User Datagram Protocol accessible for communication is an efficient, less complex application architecture in comparison to the Transmission Control Protocol (TCP). In addition, the UDP header may be compressed to make better use of the limited space available for payloading. The 6LoWPAN utilises ICMP – Internet Control Message Protocol – to monitor messages and to detect, among alia, the inaccessible destination and the surrounding discoveries. A request-response paradigm for power-low loss networks operating in constrained circumstances is provided in the Constrained Application Protocol (CoAP). The protocol enables asynchronous message transmission and HTTP mapping, allowing users to access Things Resources through HTTP Internet.

The LPWAN allows long-term communication of "things" in the field of IoT technology. It offers low-performance communication with a low bit rate, unlike a wireless broadband (WAN) network, which requires more power to function at a high data rate. The Low Power Wide Area Network (LPWAN) utilises the LoRaWAN protocol to connect between gateways and terminals allowing various communication rates throughout the battery-operated device network. In a similar manner, the Narrow Band-Internet of Things (NB-IoT) is a 3GPP protocol for communication in LPWANs. Weightless Special Interest Group (SIG) uses three separate LPWAN communication protocols, which are the following, to allow LPWAN to communicate unidirectionally, bi-directionally and in low-power modes.

The IoT has a plethora of important applications, and the most significant of them are described here.

1.2.2.1 The Smart House

Recent advancements in personal technology have opened the way to the IoT in the construction of smart houses. Similarly to smartphones, the smart house is expected to become commonplace very soon [14]. Even though houses represent a significant financial investment, smart home gadgets provide dependability, long-term durability and cost savings. Several smart home companies, like Ecobee, Nest and others have provided examples of their work, which is moving towards being normal practice in a typical house.

1.2.2.2 Smart Cities are Places Where People Live, Work and Play

A smart city integrates the functioning of many sectors such as electricity, water, transportation and so on, making it a particularly noteworthy use of the IoT among academics. The smart grid and smart transportation help to make the metropolitan environment smarter overall. The IoT will address enormous problems in order to improve the quality of life for those who live in urban settings. Sensors are used in Internet-connected apps that are available to the general public. Additionally, the sensors are capable of identifying carved problems, general failures and other infrastructural issues that may exist in the control system. Agriculture sector has grown as the demand for food has unexpectedly increased in recent years because of the continuous increase in the overall population. Governments are educating agriculturists on how to use thrust techniques and do research in order to create a more productive generation. Smart farming is one of the fastest developing areas in the IoT. Farmers are making use of an important component of the information gained from aggregated data that will assist them in increasing their production. Sensing the humidity and nutrients present in the soil will aid in the management of water and fertilizer application on a case-by-case basis, which are two of the most important objectives of the IoT in agriculture.

1.2.2.3 Retailing in the 21st Century

Nowadays, the retail business is reliant on a wide range of technologies with the IoT playing a critical role in this sector. The IoT provides the ability to link with other components and makes it simple to update and comprehend the present state. The smartphone is a critical tool for connecting with other business partners and customers, even while they are away from their physical shop locations. Collaboration via cellphones and the use of technology allows shops to operate more efficiently and provide better service to their customers as a result. The ability to monitor their customers' movements within a shop and effectively position the products that sell quickly in a more visible location to avoid rush hour crowds is a significant advantage.

1.2.2.4 The Smart Grid

Power grids that are not constrained will be both intelligent and very resilient in their operation. The aim of smart grid interconnectivity is to integrate all of the power supply lines into a single global network that is accessible from anywhere on the planet. With smart grid systems, the primary goal is to gather data in order to create an automated design that separates power users and improves both the effectiveness and the cost of energy consumption. Smart grids will also have the capability of identifying power outages as soon as they occur as well as integrating renewable energies such as solar, wind and tidal energy.

1.2.2.5 Healthcare

Despite this, concomitant healthcare insurance continues to be the finest of the IoT applications. The concept of a linked healthcare network, as well as smart wearable portable devices, has tremendous promise for both the healthcare industry and for the person. According to a recent research, the IoT will have a significant impact on healthcare in the next years. The IoT in healthcare is no longer used to encourage people to adopt a more beneficial way of life by wearing linked gadgets. The information gathered is used to assist a personalised examination of a person's well-being and to develop personalised methods for fighting illness.

1.2.2.6 Poultry and Agricultural Production

The IoT devices are being used in the poultry and agricultural industries to collect information such as health and location of domesticated animals, and these actions save money and are simple to monitor. This approach also benefits farmers by providing them with timely information on the health of their animals and by reducing the amount of money they spend on needless expenditures. Farmers may extend their poultry operations to a larger degree with the help of the data they have collected [19]. Animal health monitoring also can perform with the combination of IoT and Machine Learning technologies [20].

1.3 Challenges

The IoT ecosystem is constructed on top of a heterogeneous environment, which poses a number of difficulties, which are explored further in the subsequent sections.

1.3.1 Adding Value to the Customer's Experience

Identifying and describing different problems will guarantee that IoT deployments are completed successfully. Over the long term, companies that are focused on the IoT miss out on enormous opportunities in terms of efficacy, customer loyalty and efficiency. A detailed retrospection of the whole cycle is required in light of the present IoT use problem and knowledge of the client issue declaration.

1.3.2 Analytical Challenges

Because of the scalability and flexibility of the Cloud, IoT analytics takes place in the data centre, where it is nearly like a server-client scenario. As a result of the high latency and bandwidth needs of the Cloud, IoT analytics is not recommended. On the other hand, the identical situation occurs on the Internet of Vehicles (IoV) and commercial jets are two examples. As a result, real-time event processing utilising IoT analytics continues to be a challenge.

Currently available Cloud architecture cannot cope with the deluge of data generated by the expansion of information, gadgets and networking infrastructure. Users who can access the Cloud effectively and cheaply may take use of the vast computing, storage and even connection that the Cloud provides. In the case of devices that are situated far from a centralised Cloud, such as a data centre, these centralised resources may cause significant delays and performance degradation.

It is important to note that edge computing, also known as "Edge," performs processing close to the data source and should not be sent to the distant Cloud or other unified frameworks for handling. Reduced data transmission time to the source results in increased speed and efficiency of data transfer as a result of this improvement.

Fog computing is a standard that specifies how edge processing should be implemented and operated. The job of figuring out, stockpiling and system administration advantages across end-to-end gadgets, where data values are recognised in the Cloud, are also facilitated. Aside from that, Fog off-loading the Cloud for edge processing is applicable to a wide range of applications.

According to the following examples discovered in normal mechanical activities [18], there is a variety of essential skills that Edge may offer for Industrial IoT (IIoT) applications, taking into consideration all of the requirements of the present problems. As part of Edge, PCs and capacity frameworks are situated at the edge of the network, as near as is reasonably possible to the segment, gadget, application, or person who generates the data being managed. Because the data does not need to be transmitted, the goal is to reduce latency as much as possible, moving from the system's periphery to a central organising framework, and then back to the periphery.

The IoT-connected device is a natural application for edge processing. The use of remote sensors on a machine, either partially or completely, results in the production of enormous amounts of data. The data that is transmitted across an extended network must be analysed, recorded and tracked in order to ensure its security. This requires much more time than if the information is processed at the edge, close to the data [3], where it takes significantly less time. Using the phrase "fog processing," which was invented by Cisco, we can also refer to the process of spreading out by registering at the edge of the system. When Cisco introduced its Fog computing technology in January 2014, it was described as a method of delivering Cloud processing capacity to the system's edge. Fundamentally, Fog represents the norm, while the Edge represents the concept. In the Edge computing concept, fog enables repeated structure in order to push focused frameworks or Clouds for better and more flexible execution. To process the data present in the database or file and then analyse the data quickly, Sagar et al. proposed the new methodology [21].

1.4 Smart City Applications

Innovative ideas in academics, industry and government are emerging as a consequence of the development of intelligent cities. As long as it fosters a long-term economic development and a good quality of life while guaranteeing sensible natural resource management through participatory governance, investment in human and social capitals as well as traditional and modern communications infrastructure may enable "smart" cities. A clever city links its physical infrastructure with its ICT infrastructure, social infrastructure and economic infrastructure to draw on the collective wisdom of its people. Smart towns are frequently constructed on the basis of contemporary infrastructure and state-of-the-art technology instruments.

In the literature of digital and intellectual cities, the concept of intelligent city is examined from the point of view of technologies and materials with specific features. It seeks to achieve the most recent developments in mobile and digital computing and also to combine more recently technology with urban physical settings. A distinctive smart cities characteristic is the focus on smart embedded devices compared to smart cities. Smart cities, by contrast, create territorial innovation systems that integrate information-intensive activities, cooperative learning establishments and collective intelligence web-based apps. Addressing cities' problems and objectives in a globalisation and innovation-driven world is the major issue for the smart city environment. First of all, a dense network of broadband networks may be created to serve a broad variety of digital applications in cities that are smart. This covers: 1) the development of the cable, fibre-optics and wireless infrastructure which provides high connectivity and bandwidth for urban residents; and 2) the enhancement of cities' physical space and infrastructure with embedded systems, smart devices, sensors and actuators that deliver real-time data, alerts and a wide range of services. Open data, semantic web technology and future media technologies, as well as current cloud computing and future IoT have many potential applications to provide. Using these technologies, infrastructure scale savings, application standardisation and turn-key software services may be achieved, thereby decreasing development costs and minimising the learning curve for the administration of intelligent cities [22].

Installation of sensors and the connection to the Internet via particular information exchange and communication protocols, to enable intelligent accurate identification, tracking and management (GPS), optical character recognition (OCR) and laser scanners is a fundamental idea in IoT. A smart city needs the technological assistance of IoT devices to be instrumented, linked and intelligent. The IoT is only a few instances of smart connected homes, wearables and healthcare. It is not wrong to argue that the IoT is now a part of every field of our lives. In addition to improving the comfort of our lives with Internet applications, it also offers us with greater control over our everyday lives by facilitating routine professional and personal activities.

Due to the recent hype surrounding the IoT, companies were required to be a leader in developing key IoT building blocks including hardware, software and support to enable developers to implement applications that can connect anything within the scope of the IoT. The IoT may be utilised in a variety of applications, such as the following:

- Intelligent households
- Connected vehicles
- Industrial Internet
- Intelligent cities
- IoT in agriculture
- Smart retail
- Energy commitment
- IoT in healthcare
- IoT for poultry and agriculture

Cities are starting to face problems which impede their socio-economic and long-term development as a consequence of the increasing population and urbanisation. Complex measurement and communication technologies have allowed smart cities to play a major part in the resolution of the economic, social and institutional problems arising from industrialisation. A series of forecasting tools have been included in this context, which utilise historical statistics from the above-mentioned technologies to manage environmentally important components, such as smart grids and Smart Transport Systems (ITS) efficiently. As far as the accuracy of prediction is concerned and the administration of these intelligent systems is concerned, the data accessible in the space has now been added to the prediction tools in addition to the time data available, as was previously done. These sophisticated spatio-temporal techniques may include all available data from different places, allowing for more accurate predictions [23].

1.4.1 Challenges Faced by Smart City Applications

The challenges in attaining intelligent city status may be classified into six groups. They are categorised into several categories:

(1) Economic limits;
(2) Technological limitations;
(3) Societal limitations;
(4) Governance restrictions;
(5) Environmental restraints; and
(6) Legal and ethical limitations.

Governance problems are the plague of intelligent urban growth in the overwhelming majority of nations throughout the globe. Moreover, economic instability is a major issue. Research has shown that the long-term consequences of instable and ambiguous conditions in many economies are important, especially for long-term development, livelihood costs and infrastructure delivery. The lack of a supportive environment, prevalent in most impoverished countries, may hinder the development of clever cities. Most countries have social issues such as unemployment, lack of common cohesion, inequality and poverty as well as the lack of education chances which prevent them all from building intelligent cities. Since the commencement of the fourth industrial revolution, many industrialised and developing countries have embraced different Industry 4.0 technologies in various sectors of their economy (Industry 4.0). IoT, large data, robots, automation, drones, sensors, service digitalisation and club-based production became common technical elements that were utilised in certain nations to enhance service delivery, as shown by the adoption of these features in the United States. The arrival of wireless connection of the 5G that promises the broad-based adoption of the smart city concept was also welcomed in most countries. Another major issue is the high cost of Internet access, as well as the lack of access and bad online connection, which afflicts the majority of African countries. Persons inside an environment also play a key role in the effective implementation of smart applications. Due to population expansion, the quantity of data produced increases, which may harm the accomplishment of smart cities if it is not managed properly. In addition, population expansion leads to congestion in transit, pollution and an increase in socio-economic disparities. Lack of sustainability care (such as proper waste management) may lead to poor living conditions for city inhabitants. In addition, legal concerns are important, especially in view of the role that enormous data collection plays in intelligent city ideas [24].

One of the major problems of intelligent cities is to ensure that the required infrastructure is in place. An ideal smart city will have a broader infrastructure for public transit, healthcare and education, energy and other public services. Cities must be prepared to make use of advanced technology to effectively implement a solid plan for the growth and development of smart city infrastructure.

One of the problems facing intelligent cities is the shortage of resources. In most places, the deployment of new smart technology is extremely costly. As a result, the funds for the transformation of smart city projects are insufficient. The government may be needed to request additional money from the private sector, which is likely to be interested in these intelligent city projects to deal effectively with that problem. Intelligent initiatives in smart cities require creating new talents to be used to apply these technologies [25]. Big data integration with IoT may assist the creators of intelligent cities to solve their technological problems.

1.4.2 The Necessity of Integrating Big Data and IoT

Various designs for the big data analysis and IoT were proposed to deal with problems of storing and analysing large amounts of data from intelligent buildings. The first organisation mentioned has three main components: Big Data Administration, the IoT sensors and the analysis of data utilised for real-time oxygen management, risk levels of gas and the quantity of ambient illumination in smart buildings. In addition to the intelligent management of buildings, traffic information from IoT devices and sensors may be utilised at low cost in real time and analysis of the strengths and weaknesses of existing traffic systems can be carried out.

In smart city management, big data is utilised for the evaluation of data from various sensors, including water sensors, transportation network sensors, monitoring devices, smart home sensors and smart parking sensors. These data are produced and analysed in a multi-stage model with the final stage of decision-making. Process stages include data creation, data gathering, data integration, data segmentation, data processing and decision-making.

In order to be successful, a number of IoT ideas must be addressed at certain moments, in particular, a suggested framework to analyse the analytical results of massive data collected through the IoT. In the literature [26], this issue was addressed and a five-level conceptual framework was suggested for solving it:

- Data collection layer – data collected from a number of sources; the input layer is the provided framework.

    The load for extract transformation (ETL) in the sensor network may convert information gathered from different sensor types into a predefined format.

    It consists of an implication engine, which works on the information received from the ETL layer and the semantic reasoning rules.

    This layer pulls from data that are customised to fit existing extraction data the various needs and features. Finally, it offers machine-based models.
- Action layer – this layer executes a number of predefined actions using the outputs of the learning layer.

1.4.3 To Use IoT with Big Data to Solve the Problems of Intelligent Cities

The integration of IoT measuring technologies with big data analytics will provide a new approach to solve the challenges of intelligent cities. Integration would be advantageous in many sectors, including water, energy, agriculture and others. ITS are a key part of smart cities and are becoming more popular. Generally, ITS have a three-tier architecture which includes users, vehicles and roadways, all of which function together as a service network. There are many transport components in cities, such as real-time traffic information, group transit, speed limitations, emergency vehicle assistance, etc. Examples of the following are: A smart city coordinates and implements all these kinds of mobility so that the social problems caused by traffic congestions are kept to a minimum. The advantage of ITS is thus a decrease in traffic congestion and, therefore, a decrease in fuel consumption, and, in particular, life-saving by reducing road accidents and by providing efficient vehicle emergency assistance [19, 27].

Big data, produced by IoT to power smart cities, has an effect on policy choices and urban government. This data, collected and analysed at rates that were previously unimaginable for people decades ago, has an impact on urban governance and policy decisions. This opens up new methods of evaluating how cities function economically, healthily and well-being. Despite the increasing digitisation of cities through a process accelerated by the Smart Cities concept, the interpretation of sensor data from the IoT equipment restricted the exploration of urban health, except data relating to human anatomy and the development of biological data in various forms [28, 29].

The recent developments in IoT and big data technology have given a broad range of opportunities for developing disaster-flexible intelligent city settings that are much needed in today's society [29]. The integration of IoT and BDA technologies has produced a new Reference Design and Concept for a Disaster Resilient Smart City (DRSC), and it was released. This architecture offers a generic approach to catastrophe management operations in intelligent city incentives, which may be applied across the board. The Hadoop Ecosystem and Spark are used to create an organised DRSC environment that supports both off-line and in real-time analysis. Data collecting, data aggregation, pre-processing data, big data analytics and service platform components constitute the implementation model of the environment. In order to validate and evaluate the system's ability to detect and produce building fire warnings; urban pollution levels; emergency evacuation routes; and natural disaster data sets; various data sets are used, such as town pollution, Twitter and traffic simulators in the event of tsunamis and earthquakes.

Continuous monitoring and evaluation of environmental factors will generate huge quantities of information that will need scientific and technological innovation to build a sustainable management plan. A framework has been developed to handle six environmental issues which should be examined and addressed together in smart city building. These ecological components consist of landscape indicators and geography as an important environmental component, climate and pollution of air, water and energy resources and urban green space. These landscape and geographical indicators are also covered by environmental variables. The approach developed is based on large-scale data analysis [30].

A lot of what is said about Big Data has been spoken already, and it has been expressed many times by people who are already familiar with the matter. With access to this data volume, researchers may evaluate Big Data's "potential" in terms of the delivery of services and policy formation. This covers issues such as data quality and reliability via the mixture of public and private sector data, raw and modified ownership of data and ethical considerations about surveillance and privacy protection. These results and the problems mentioned help evaluate the value of large data along with IoT in government and municipal contexts [31, 32]. These findings are also highlighted.

In order to create an objective function-based fuzzy mean clustering algorithm theory utilising big data analysis technology, K-means and fuzzy theory are merged into big data analytics technology. This algorithm idea is used by large data analysis technologies to investigate electric vehicle networks in intelligent cities (FCM).

According to results from a more thorough study of the effect of different variables on transport conditions, such as the market penetration rate for equipment (MPR), the following rate (FR) for vehicles and the development of congestion (CL), the induction strategy is improved. Big data analysis technology is used to improve the functioning of transmission networks for electric cars. This technology has the potential to significantly decrease the network data transfer latency and change the route to manage congestion spread efficiently, and it has served as a test bed to build electric vehicle transport networks [32].

Token-based framework use for IoT devices concept used for accessing the cloud services is also good method. Kafka can also use for streaming of health data [33, 34]. S., Motupalli, R., et al. applied the IoT and Block Chain Technologies for smart transportation for smart city world [35, 36].

In addition, many more areas of intelligent cities are developed with the integration of IoT and Big Data. One such example is described briefly in this section. The example of an IoT and Big Data integration fire service that overcomes obstacles is taken into account in this regard.

1.5 Case Study

IoT and big data integration in smart homes Fire Department Services: This case study shows in what way a city that gathers and distributes its data to the public may improve the sustainability of its activities via evidence gathered using big data techniques and visualisations. The data supply procedure may be carried out via a number of sources. For one instance, we have developed the benefit of Exposed Information gateways. The process for dealing with Open Data is as easy as it is for working with any other data source and there is no limit on its use or publication. In the event of open information, "stuff that may be freely used, modified, and shared for any purpose by anybody," according to the definition has been described. Consequently, we have utilised the publicly accessible San Francisco data collection as our input. One specific dataset utilised in this research was the Calls for Service dataset for the Fire Department. By April 2020, the dataset consists of 5.27 million rows and 34 columns; each row represents a unit request. This city would need a collection of visualisations in order to analyse its data to increase the response time of its emergency services in order to boost its long-term sustainability. We pretend to be a user that uses available resources to minimise the harshness and number of important fires in the metropolitan. According to Figure 1.6, the suggested approach is accessible via the integrated process for this case study.

Then, in this research, the user decided in this study for the creation of two distinct kinds of analysis. A "prescriptive research" to examine the historical data and decide how the city may become a more dependable Clever City. The operator too creates a "predictive analysis" to anticipate what will happen in order to be able to respond more effectively. Each kind of analysis is split into a set of decision goals. Decision objectives are aimed at motivating individuals to take appropriate action to achieve a planned box and explain how they may be achieved. For the prescription analysis, users set the decision goals as "Identify fire hazard" and "Analyse battalion performance." In addition to defining one or more information goals for each decision objective in order to offer more detailed information, the user may set one or more decision goals for each information objective. The lowest degree of objective abstraction is the information objectives. In order to achieve the "fire

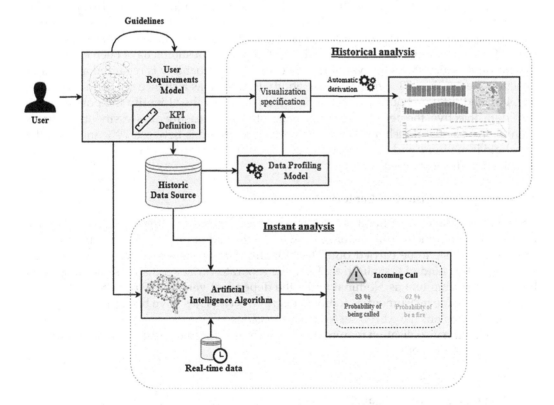

FIGURE 1.6
Integrated fire department framework.

risk identification" decision objective, users must first select the information objectives of "identifying month with the most fires," "identify week days with the most fires," "identification of the most fire times" and "identifying districts with the most flames." Besides this, to achieve "Analysis of Battalion Performance," the user has selected two information goals: "Battalion average time of arrival" and "Number of calls replied by Battalion each year." In order to carry out prognostic examination, the operator describes "Reduce arrival time" as the choice objective which is further specified by the information objectives "Identifying the probability of action by the unit" and "Identifying fire chance." To accomplish each information objective, a visualisation will be produced automatically (the visualisations used to attain the material goals "Identify month with more fires," "Identify days of the week with more fires," and "Identify time slot with more fires" all follow the same requirements, which is why they have been gathered together in the model). A visualisation is characterised by single or more visualisation goals and several interactions. The visualisation goals specify the components of the visualisation data, and the interaction type indicates the users' desire to participate in the visualisation. The visualisation objectives are further explained below. The recommendations given may be utilised by users to describe these components. In this specific case, the operator selected "Trend," "Distribution," "Geospace," "Comparison" then "Composition" visualisation goals. In addition, the "Overview" and "Details on Demand" communication types are provided.

Finally, before finishing the selection process, the operator chooses the analytical data source, as well as the categories and measures to supplement the views. After a User Requirements Model has been created, users will define the key performance indicators (KPIs). In this instance, the number of severe fires was designated as a critical KPI for analysis of the strategic objective of "Reduce major fires." It is recommended that a percentage be used to estimate the chance that the unit is called to accomplish the "Identify unit action probability" information goal. The following thresholds have been established to determine which states this variable may take: High probability is regarded as a chance of action greater than 70%, medium probability is defined as a chance of action amid 70% and 40% and little probability of action less than 40%. For the "Identify the potential of a fire" data objective, the percentage is also suggested as a measure of assessment. If a fire is more than 70%, the threshold is high; when the likelihood is between 70% and 40%, the threshold is medium; and the probability less than 40%, the threshold is low. The threshold is defined in this instance as follows:

1.5.1 Analyse the Historical Context (Prescriptive)

The following stage is to put on our data summarising perfect for the selected data source in order to determine its dimension, cardinality and kind of data. This is essential for the historical analysis to be carried out. The "Identify Fire-prone Areas" visualisation is an example of this kind of visualisation. First, the separate variable "districts" is categorised by a data profiling tool as Nominal while the dependent variable "Fire quantity" is classified by the same instrument as Ratio. Because two variables have been provided by the user, the dimensions have been set to two dimensions.

The following visualisation requirements have been created using the Operator Supplies and Data Profiling models:

- Visualisation goals: distribution
- Imagining goals: geospatial
- Interaction: User overview

- Lay
- Size: 2D
- Cardinality: in height
- Independent: insignificant
- Independent: relation
- Independent: Category
- Independent: Nominal

A "MAP" is a graph better suited to this standard. This same process is used for each of the other visuals forming the model to decide which kind of visualisation is best suited to each given specification.

Predictive analysis is a type of future analysis. However, historical analysis may not always offer all the necessary information for decision-making. Predictive analysis will be conducted in the actual time situation, in which the operator needs to forecast "probability of unit action" and "probability of flames." In order to anticipate the future, an artificial intelligence algorithm will be integrated into the system. When a call is received, this algorithm will analyse it and create a prediction before responding it, in order to provide a warning when the battalions can fight.

The Historic Data Source is utilised for the training of artificial intelligence. This algorithm will find out which elements in the data stream are to be used to determine the battalion's call type and probability of action. The user has chosen the whole source for feeding visualisations from Predictive Analysis in the User Requirements Model, which is why we utilise the complete data source. The Deep Neural Network (DNN) is six layers dense with 256 neurons (input, output and four hidden ones). To prevent data bias issues, the first step is to change and split the whole dataset into three groups: train data, validation data and examination information. In this instance, 70% of the information is used to experiment the algorithm, 20% to verify the algorithm and 10% to test it. The algorithm is provided with train and validation data, and this data will be second hand by the procedure to study in what way to classify. The test results are then used to evaluate the learning result. A layer of 64 nerve cells is then created for each variable and initiated. Then these neurons will be combined with a layer of 32 nerve cells using a random number generator. Finally, these neurons collectively merge into a set of 96 neurons, which are spitted 64 neurons, shadowed by 32 neurons, and as a result the output is a second nerve cell. This is the most difficult job and the greatest processing power is required. Once fresh data are introduced into the procedure, such as a novel noise, the procedure can automatically and quickly anticipate the process information once it is trained and evaluated.

Once this method is complete, it provides a value between zero and one, near zero, suggesting that here is no passion hazard and a near to one indicating that nearby is considerable passion risks. These visualisations are composed of possessions that stipulate the verge values for each view. These requirements have remained laid out in the past. As a result, these visualisations will take these specific criteria into account and enable visitors to see the algorithm results. Besides the visual representation of the probability of action, the "Send Battalions warning" work will be performed to fulfil the objective of the decision "Reduce arrival time" and "Reduce arrival time." This means that those battalions with a higher probability are informed in advance. As a result, they may be prepared for the action, and the decrease in arrival time in turn will result in a decrease in fire impact. During incoming calls, customers get instantly a new message that they haven't received earlier. This has been achieved with the artificial intelligence and the results also were made to be accurate. It also provides visual ways to assist users to effortlessly appreciate the procedure yield and the capacity to establish verges to distinguish between the output and the procedure. These visualisations are collected into a dashboard which is refreshed at every call and ensures customers have access to the latest information.

1.6 Final Dashboard

An interactive dashboard is created after all the visualisations have been completed. The dashboard brings together all the visuals that have been produced in order to meet the analytic needs of our fire department supervisor user. Using the User Requirements Model, the visualisations that make up the dashboard are organised into categories based on the kinds of analysis that are performed. The Service Calls Management process is represented by this dashboard, which offers an overview of the process. The historical evaluation is displayed on the left-hand side of the screen. Using this tool, the user may determine the level of fire danger at a certain point in time, as well as how fires are distributed throughout various districts. The operator whitethorn also assesses the presentation of the Battalions by looking at data such as the average time required to arrive at a division and the amount of demands responded by Battalion, among others. On the contrary, the vital metrics will be shown on the right-hand side of the screen, making it easy to evaluate each incoming call before answering it. As calls come in, these visualisations will be updated to reflect the likelihood of a fire as well as the likelihood of the Battalions being sent. The last important indicator is located at the top of the dashboard and measures the strategic objective of "Reduce severe fires." It is shown in red. With the use of this KPI, users will be able to track the effect of their choices and determine whether or not the measures in place are really increasing the city's sustainability.

1.7 Conclusion

Smart cities have become a need in order to solve the problems that have arisen as a result of increasing urbanisation. The solutions presented in this chapter show how cities have addressed these problems in order to improve the overall life quality for its residents and visitors. There is a rapid increase in the number of cities globally that are seeking smart transformation. But there are many hurdles in the way of these endeavours in the political, economic and technological spheres. It is necessary to take into account a number of variables and difficulties before proceeding forward. Many times, smart city projects require extensive coordination, financing and on-going assistance. It is necessary to provide a return on investment, which poses still another difficulty. Another important factor in ensuring security and privacy is the presence of technological barriers. Additionally, tolerating a proliferation of resources and infrastructures is very essential for any smart city efforts in the long term. Because the deployment of IoT infrastructures may open the door to a plethora of new and exciting possibilities; the highest research motives are first described, followed by a discussion of specific practical applications. It describes how their use may help to grow, improve and enhance everyday tasks in many ways. Incorporating the IoT with big data analytics offers a potential answer for the problems encountered by smart cities in terms of infrastructure for computation, organisation for storage, data diversity, data-study and statistics performance for the benefit of the bookworms.

References

1. P. Yadav and S. Vishwakarma, "Application of Internet of Things and big data towards a smart city," in *2018 3rd International Conference On Internet of Things: Smart Innovation and Usages (IoT-SIU)*, 2018, pp. 1–5.
2. A. Kirimtat, O. Krejcar, A. Kertesz, and M. F. Tasgetiren, "Future trends and current state of smart city concepts: A survey," *IEEE Access*, vol. 8, pp. 86448–86467, 2020.
3. B. Silva, M. Khan, C. Jung, J. Seo, D. Muhammad, J. Han, et al., "Urban planning and smart city decision management empowered by real-time data processing using big data analytics," *Sensors*, vol. 18, p. 2994, 2018.
4. M. N. I. Sarker, M. N. Khatun, G. M. M. Alam, and M. S. Islam, "Big data driven smart city: Way to smart city governance," In *2020 International Conference on Computing and Information Technology (ICCIT-1441)*. pp. 1–8, 2020, September, IEEE.
5. N. Sudhakar Yadav, K. G. Srinivasa, and B. Eswara Reddy. "An IoT-based framework for health monitoring systems: A case study approach," *International Journal of Fog Computing (IJFC)*, vol. 2 (1), pp. 43–60, 2019.
6. Sudhakar Yadav, B. Eswara Reddy, and K. G. Srinivasa. "A centralized health monitoring system using health related sensors integrated to the mobile," *International Journal of Vehicular Telematics and Infotainment Systems (IJVTIS)*, vol. 2 (1), pp. 68–79, 2018.
7. R. Casado-Vara, P. Novais, A. B. Gil, J. Prieto, and J. M. Corchado, "Distributed continuous-time fault estimation control for multiple devices in IoT networks," *IEEE Access*, vol. 7, pp. 11972–11984, 2019.
8. M. Stoyanova, Y. Nikoloudakis, S. Panagiotakis, E. Pallis, and E. K. Markakis, "A survey on the internet of things (IoT) forensics: Challenges, approaches, and open issues," *IEEE Communications Surveys & Tutorials*, vol. 22, pp. 1191–1221, 2020.
9. A. A. Mirani, M. S. Memon, M. A. Rahu, M. N. Bhatti, and U. R. Shaikh, "A review of agro-industry in IoT: Applications and challenges," *Quaid-E-Awam University Research Journal of Engineering, Science & Technology, Nawabshah*, vol. 17, pp. 28–33, 2019.
10. S. Kumar and M. Singh, "Big data analytics for healthcare industry: Impact, applications, and tools," *Big Data Mining and Analytics*, vol. 2, pp. 48–57, 2018.
11. F. Ghofrani, Q. He, R. M. Goverde, and X. Liu, "Recent applications of big data analytics in railway transportation systems: A survey," *Transportation Research Part C: Emerging Technologies*, vol. 90, pp. 226–246, 2018.
12. L. Zhu, F. R. Yu, Y. Wang, B. Ning, and T. Tang, "Big data analytics in intelligent transportation systems: A survey," *IEEE Transactions on Intelligent Transportation Systems*, vol. 20, pp. 383–398, 2018.
13. A. M. S. Osman, "A novel big data analytics framework for smart cities," *Future Generation Computer Systems*, vol. 91, pp. 620–633, 2019.
14. Y.-S. Jeong and J. H. Park, "IoT and smart city technology: Challenges, opportunities, and solutions," *Journal of Information Processing Systems*, vol. 15, pp. 233–238, 2019.
15. M. Gheisari, G. Wang, W. Z. Khan, and C. Fernández-Campusano, "A context-aware privacy-preserving method for IoT-based smart city using software defined networking," *Computers & Security*, vol. 87, p. 101470, 2019.
16. R. H. Hariri, E. M. Fredericks, and K. M. Bowers, "Uncertainty in big data analytics: Survey, opportunities, and challenges," *Journal of Big Data*, vol. 6, pp. 1–16, 2019.

17. C. D. G. Romero, J. K. D. Barriga, and J. I. R. Molano, "Big data meaning in the architecture of IoT for smart cities," in *International Conference on Data Mining and Big Data*, 2016, pp. 457–465.
18. S. Shadroo and A. M. Rahmani, "Systematic survey of big data and data mining in internet of things," *Computer Networks*, vol. 139, pp. 19–47, 2018.
19. N. Sudhakar Yadav, B. Eswara Reddy, and K. G. Srinivasa, "An efficient sensor integrated model for hosting real-time data monitoring applications on cloud." *International Journal of Autonomic Computing*, vol. 3 (1), pp. 18–33, 2018.
20. N. S. Yadav, M. P. B. Reddy, and G. Sreenivasulu, "ML and IoT based real-time health monitoring system for domestic animals." *Journal of Critical Reviews*, vol. 7 (19), pp. 10111–10117, 2020.
21. N. Sudhakar Yadav, S. Yeruva, T. S. Kumar and T. Susan, "The improved effectual data processing in big data executing map reduce frame work," in *2021 IEEE Mysore Sub Section International Conference (MysuruCon)*, 2021, pp. 587–595, doi: 10.1109/MysuruCon52639.2021.9641660.
22. A. Das, S. C. M. Sharma, and B. K. Ratha, "The new era of smart cities, from the perspective of the internet of things," in *Smart Cities Cybersecurity and Privacy*, Elsevier, 2019, pp. 1–9.
23. A. Tascikaraoglu, "Evaluation of spatio-temporal forecasting methods in various smart city applications," *Renewable and Sustainable Energy Reviews*, vol. 82, pp. 424–435, 2018.
24. D. O. Aghimien, C. Aigbavboa, D. J. Edwards, A.-M. Mahamadu, P. Olomolaiye, H. Nash, et al., "A fuzzy synthetic evaluation of the challenges of smart city development in developing countries," *Smart and Sustainable Built Environment*, 2020.
25. E. Okai, X. Feng, and P. Sant, "Smart cities survey," in *2018 IEEE 20th International Conference on High Performance Computing and Communications; IEEE 16th International Conference on Smart City; IEEE 4th International Conference on Data Science and Systems (HPCC/SmartCity/DSS)*, 2018, pp. 1726–1730.
26. Z. Alansari, N. B. Anuar, A. Kamsin, S. Soomro, M. R. Belgaum, M. H. Miraz, et al., "Challenges of internet of things and big data integration," in *International Conference for Emerging Technologies in Computing*, 2018, pp. 47–55.
27. M. Gohar, M. Muzammal, and A. U. Rahman, "SMART TSS: Defining transportation system behavior using big data analytics in smart cities," *Sustainable Cities and Society*, vol. 41, pp. 114–119, 2018.
28. Z. Allam, H. Tegally, and M. Thondoo, "Redefining the use of big data in urban health for increased liveability in smart cities," *Smart Cities*, vol. 2, pp. 259–268, 2019.
29. S. A. Shah, D. Z. Seker, M. M. Rathore, S. Hameed, S. B. Yahia, and D. Draheim, "Towards disaster resilient smart cities: Can internet of things and big data analytics be the game changers?," *IEEE Access*, vol. 7, pp. 91885–91903, 2019.
30. R. Dwevedi, V. Krishna, and A. Kumar, "Environment and big data: role in smart cities of India," *Resources*, vol. 7, p. 64, 2018.
31. K. Löfgren and C. W. R. Webster, "The value of Big Data in government: The case of 'smart cities'," *Big Data & Society*, vol. 7, p. 2053951720912775, 2020.
32. Z. Lv, L. Qiao, K. Cai, and Q. Wang, "Big data analysis technology for electric vehicle networks in smart cities," *IEEE Transactions on Intelligent Transportation Systems*, vol. 22, pp. 1807–1816, 2020.
33. N. Sudhakar Yadav, et al., "Accessing cloud services using token based framework for IoT devices," *Webology*, vol. 18 (2), 2021.
34. N. Sudhakar Yadav, B. Eswara Reddy, and K. G. Srinivasa, "Cloud-based healthcare monitoring system using Storm and Kafka," in *Towards Extensible and Adaptable Methods in Computing*, Springer, Singapore, 2018, pp. 99–106.
35. R. Motupalli, et al., "An automated rescue and service system with route deviation using IoT and blockchain technologies," in *2021 IEEE Mysore Sub Section International Conference (MysuruCon)*, IEEE, 2021, October, pp. 582–586.
36. Chalumuru Suresh, et al., "Cognitive IoT-based smart fitness diagnosis and recommendation system using a three-dimensional CNN with hierarchical particle swarm optimization," *Smart Sensors for Industrial Internet of Things*, Springer, Cham, 2021, pp. 147–160.

2

Challenges and Roles of IoT and Big Data Analytics-Enabled Services in the Establishment of Smart Cities

J. Jeya Bharathi and S. S. Aravinth
Koneru Lakshmaiah Education Foundation, Vaddeswaram, India

S. Madhusudhanan
Prathyusha Engineering College, Thiruvallur, India

U. Rahamathunnisa
Vellore Institute of Technology, Vellore, India

N. Sudhakar Yadav
VNR Vignana Jyothi Institute of Engineering and Technology, Hyderabad, India

CONTENTS

2.1 Introduction 24
2.2 Related Work 25
    2.2.1 Birth of IoT 25
    2.2.2 The Internet of Things 26
    2.2.3 The Cloud Computing 28
    2.2.4 Big Data Analytics 30
    2.2.5 Opportunities and Difficulties in Big Data Analytics........................32
2.3 Efficacy in Relation to Time...32
2.4 Why Cloud of Things 33
2.5 Integration Involving Cloud and Internet Associated with Things 33
    2.5.1 A Couple of Cloud-IoT Applications 34
2.6 The "QoS" on Integration of Cloud and IoT..35
2.7 The Benefits of Converging IoT-Cloud 37
2.8 One Cloud-Based IoT Architecture 38
2.9 Difficulties in Converging IoT-Cloud 38
2.10 Conclusion 39
References..40

DOI: 10.1201/9781003217404-2

2.1 Introduction

It is crucial to study the common technical components that frame the subject of computer. Indeed, that is usually genuinely the case together with "Cloud Computing" and Web of Things "IoT", two principles that discuss several unusual characteristics. The combination of this even more than one idea can assist and beautify this period. The Cloud computing has reformed distinguish the way technology are frequently acquired, managed and distributed [1]. Cloud computing is widely acknowledged for upcoming services. Although many viewpoints Cloud computing as a replacement era, it truly is, inside fact, worried and combines various technologies like main grid, application computing virtualization, network and records offerings [2]. Cloud computing gives offerings that enable them to disperse pc resources over the net. As it will be, it is far from sudden that typically the origins of Cloud era live the grid, computer usage, communication offerings and software offerings, also for distribution computers and as a result the same pcs [3].

Alternatively, IoT are frequently regarded as a robust infrastructure and a worldwide community regarding self-regulating gadgets throughout a really smart manner [4]. It will be handed over to a level where the entirety about us is attached to the web and capable of communicating with minimum human effort. IoT typically consists of multiple gadgets together with confined storage and computer capacity. Well, Cloud computer and IoT are the particular long run of the web and accordingly the following generation of generation. Nevertheless, Cloud services agree together with interchangeable carrier providers, from the same time as IoT technologies believe diversity in place of interdependence.

Cloud computing and IoT are two different procedures, and although they each have their own constitution and specific terminology, these are very much connected with each other within our daily lives [5]. Typically, the amalgamation of Cloud and IoT is known as Cloud-IoT imagery [6, 7]. By integrating these two technologies into the so-called Cloud-IoT, we hope to fragment the current and future Internet.

In research, industry or education, adding Cloud and IoT is a really important, exciting and promising aspect. We are going to definitely explore the value of the incorporation of typically the Cloud and the IoT and discuss what problems we have to incorporate them and the way to meet these people in real time.

We are also able to easily explore the normal features and benefits of integration to integrate the particular Cloud and the Internet of Things (IoT). Not everything is cheap for everyone and not free for everybody. If you are a good Internet user, you ought to buy everything because of it. When it comes to IoT devices and Cloud servers, it is not simple for anyone to bear it. But using open resources is very good for us. In this chapter, we use mostly open supply ones. We use a new non-expensive IoT device in order to integrate IoT and Cloud servers and use Glowing Blue Virtual Machine as a new Cloud that is totally available as a college student version and open supply protocol is self-governing associated. . Usually appears like when we are transmitting text message over the Internet. The communicated message is easy for the intruder to be able to catch and it is usually not good for both the sender and the beneficiary.

Any of your personal information or even message is incredibly confidential value to secrecy. As a result of security functions, it is quite much risky. Regarding that you need to transfer the information that is secure [8, 9]. Here we all use a specific protocol term MQTT as the result of security advantages. First, we encode the complete text or string and after that pass it through the MQTT protocol. By this particular time, if any intruder can take probabilities to view a text file, they won't see it even though they are encrypted.

Your document is very secure. When typically, the text is moved to be able to the desired location, typically the receiver can decode the particular encrypted file and acquire the information securely.

Protection build-up is done extremely conveniently making use of the MQTT protocol. Here, we transfer the particular image via the MQTT protocol and we employ facial recognition as an application. When we are using a great IoT device and shifting images in typically the protocol to the remote server, it is very complicated and can cause an error. Otherwise, it could easily send any type of video file along with any size or virtually any other file as opposed to moving the image file. In addition, it gives you the choice of transferring the file securely.

Big Data analysis has come to the top of the list of issues of interest for professionals and academia in recent years. The rapid expansion of digital devices, Internet usage, tablet and smartphone use and other types of data generation generate massive amounts of data all the time. Big Data, in contrast to conventional data, is derived from a range of sources and presented in a variety of ways. The volume, diversity and speed with which this data is generated provide significant difficulties for data center managers. Computer, storage and analytical capabilities, on the other hand, have addressed these issues. Large datasets may now be stored in a simple and cost-effective manner. Given the peculiarities of Big Data, particular consideration must be given to it throughout the data extraction and ingestion process from a variety of sources. Any framework should be capable of efficiently processing massive amounts of data, as well as assimilating and processing organized, semistructured and unstructured data. Big Data were characterized in many ways. Big Data define datasets as those that are too big for conventional data processing methods and thus need the management of new technologies to be effective. The term Big Data is defined as applications that need sophisticated and one-of-a-kind storage, administration, analysis and visualization technologies in order to deal with such large and complicated datasets and analytics technology. Big Data is not only defined by its amount but also by its rapid speed, diversity, breadth and connection with other things in the natural world. In a nutshell, Big Data refers to large amounts of data.

2.2 Related Work

2.2.1 Birth of IoT

Although the term Internet of Things is just16 years old, the original concept of connected devices dates back to at least the 1970s. Previously, this supposed to be often cited as "embedded Internet" or "significant computing". Initially, "Internet of Things" became utilized by Kevin Ashton in his work on Proctor Gamble in 1999. Ashton, who works in supply chain optimization, wants to draw the eye of senior management in the direction of a new exciting technologies referred to as RFID. Since the Internet became the brand-new trend in 1999, he referred to his show as "The Web of Things". Several milestones within the improvements regarding bodily mashing with electronic digital are as follows:

a. On January 13, 1946, Dick Tracy and members of the constabularies force wore a wristwatch two-way hand radio that first deemed and became one in every of the highest recognizable symbols of the particular comedian strip.

b. In 1949, the bar code emerged while 27-year-antique Norman Frederick Woodland of Miami Seashore drew 4 drops of sand. Woodland, later a good IBM engineer, acquired the main patent in 1952 for that linear bar code. A lot more than 20 years later, any other IBM, George Lohrer, changed into chiefly accountable for improving the idea for using supermarkets.

c. Edward O. Thorp builds the first wearable computer in 1955. Thoreau, a cigarette pack-formed analog device, was used for the sole purpose of examining roulette cycles.

d. On March 4, 1960, Morton Heilig was patented regarding head-mounted performance.

e. In 1967, Hubert Upton invented an analog wearable laptop having a mirror show to help study lips.

f. On October 29, 1969, the 1st message was sent in order to the Internet's predecessor, ARPANET.

g. On June 26, 1974, the Universal Product Program Code (UPC) label changed into very first used to shop in supermarkets.

h. In 1977, Sisi Collins assisted the aesthetically impaired, a five-pound wearable with a head-hooked upwards camera that transforms snapshots on a t-shirt into a tactile main grid.

i. In 1980s, those of the Carnegie-Mellon Computer Science Department set up "micro-switches" inside the coke vending machine and connected to the PDP-10 department pc to see how many flasks were in their own device terminals and whether or not they have been cold.

j. In 1990, Olivetti advanced an active logo gadget that uses infrared indicators to talk a new person's location.

k. In 1991, Mark Weiser of Xerox PARC published "The Computer in the 20 or so first Century" in Technological American, using the words "ubiquitous computing" and "embodied vitality" to describe their vision. Specific factors of hardware and software program are related to this.

l. In 1993, MIT's Thud Starr commenced using particularly tough computers and head-up presentations as wearables.

2.2.2 The Internet of Things

In a nutshell, the IoT gives users an unparalleled way to communicate with the Web through the Internet. IoT is bringing a massive upheaval to the fast-growing telecommunications sector, primarily in wireless communication. The IoT footer is made up of self-configuring and rational content that is connected to subsets of networks all around the world.

We can simply assess and examine everything in the world, allowing us to cut costs, losses, waste, disaster and so on. We can deliver things that are more important than our priorities by integrating intelligent components with an IoT device. The IoT has evolved in several facets of life since the advent of techniques of artificial intelligence, and this piqued the interest of academics who are attempting to create a new paradigm of living norm [10]. This invention has been widely embraced across the world as a means of making life simpler as a result of the rapid proliferation and widespread usage of different smart devices such as sensors, actuators and a variety of other gadgets. The use of AI-enabled gadgets makes them more intelligent and capable of performing a particular job, which saves a significant amount of time and money. Because of their low cost and adaptability, the IoT, mobile and network applications offer a superior solution. The IoT's primary role is to connect users to readily available resources while also providing them with dependability, effectiveness and smart service. When it comes to smartness, the IoT puts sensors with seamless operations, a remote server and a network together in one package. The plan is extensive in terms of providing monitoring with multidimensional structures and delivering core therapy suggestions, among other things. The IoT takes a range of significant applications in real life that make living simpler. IoT with the integration of Big Data can resolve many issues [11, 12].

The IoT is a network through which all kinds of physical objects, vehicles, buildings, etc., can be easily connected through a variety of software, electronic devices and sensors that collect and forward data. Collecting data to measure the performance of the device is searching for the IoT, and what is possible in terms of the transmission and its safety depends on it to learn about the "IoT"; at first, we need to know about the application that needs to understand about these technologies.

Thermos tones, smart automobiles, smart cities, smart hospitals, a wide range of electronic applications, fire alarm clocks and other IoT gadgets can be found in large quantities [13, 5, 14]. Now we'll discuss how these apps can make it simple to operate any IoT device. Assume you're out of the house for a while and then return home. Given the current weather, if the sun is shining, open the fan windows or close the window with the fan if the sun is not shining. How are you going to do it? With the help of some IoT devices, you can achieve this quickly and easily. Those who utilize the Internet can simply operate a smart hospital with the same technology.

Figure 2.1 illustrates the extent of typically the IoT engagement in all household needs and that everything in your home has been managed slightly over the Internet. Thermos tones, clever cars, smart cities, wise hospitals, a variety associated with electronic applications, noisy sensors for fire signals, and so forth are located in bulk for IoT devices. Now, we're speaking about how these apps may help you easily use any IoT device. Suppose you are usually away from home and come to residence after some time. Taking a look at the weather now, the particular windows of your home, the fan windows wide open or the window combined with the fan is closed in case the sun is shining. How will you do this? You may do this easily by using some IoT devices involving the Internet.

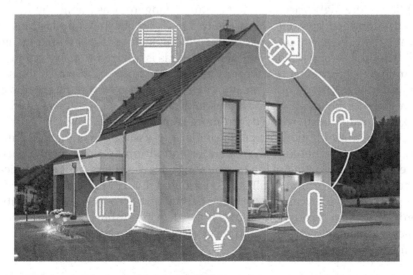

FIGURE 2.1
Smart home.

2.2.3 The Cloud Computing

Hinder computing permits convenient, on-call for and scalable neighborhood get entry to the particular configured computing source pool. Cloud computing provides inside reality unlimited talents within terms of garage area along with processing power. Some terms related to Cloud computing are explained below:

a. Cloud computer is a type for enabling on-demand group acquire right of admittance to be capable in order to a shared pool associated with configurable resources (including systems, servers, garage, packages additionally services) with minimal effort or speedy services provider interaction. Issued. [National Institute of Standards and Technology (NIST)]

b. Use of technological developments for external clients is largely a fragmented work style, with mostly global IT-enabled abilities (Gartner).

c. C. When compared to the Internet, the expanding IT development, application, and distribution paradigm enables more timely delivery of goods, providers, and solutions."Cloud computing" describes a solution model, that is IT delivery, infrastructure components, a framework and primarily economic building, grid computing as typically the service, virtualization, utility computer system, hosting and Software as a Service (SaaS). However, each research class defines Cloud computing coming from its attitude, making it tough to be able to define a much more general explanation. Cloud computing is typically classified in the following four ways:

d. Public Haze up computing infrastructure will be organized by the way involving typically the Cloud seller inside the general public Cloud. The consumer does not necessarily have visibility Considering that manner in which the conventional computing facilities are run. The computer infrastructure will be shared among any corporation.

e. Private Cloud: The processing infrastructure is committed to a selected corporation and can't be shared along using other groups. Some specialists believe that private ambiance aren't real examples regarding Cloud computing. Private Cloud is incredibly more high-priced and not as much dangerous as public Cloud.

f. Hybrid Cloud corporations can easily host complicated applications inside typically the private fog up and extremely little defense issues within the general public Cloud. Using personal and public Clouds is referred to as hybrid impair.

g. Community fog up: This is a form of fog up web hosting, in which the setup is reviewed in turn among quite an amount of businesses within the specific group, which includes financial organizations and merchants. It will be a multi-tenant setup. This really is shared between more compared with how one entities of a new particular group, which include similar processing fears. Fog up computing is the disruptive technology that offers profound implications for the particular whole IT sector, consisting of Internet offerings. Nevertheless, a lot of technical and business-related difficulties still stay unresolved. Every single service model has exact problems, which can end up being particularly associated with security (e.g., records protection throughout addition to integrity, system protection), personal privacy (e.g., statistics privacy) and service-stage agreements, which may possibly intimidate potential customers.

A few types of Cloud works are usually summarized and shown in Figure 2.2. Fog up computing uses the electricity paradigm to produce and consume resources for computation,

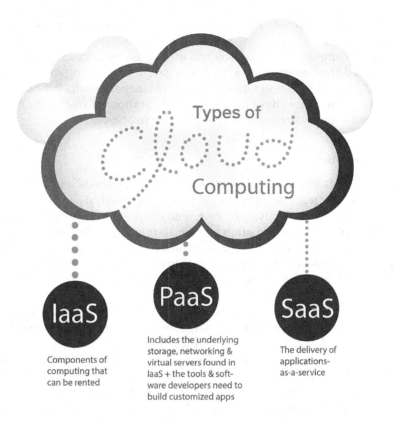

FIGURE 2.2
Cloud computing categories.

whereby the Cloud stores almost all varieties of computing resources (along with Cloud services). Cloud customers (utility building contractors or application customers) may access Cloud services more than the Internet and hinder users pay handiest with regard to that time and providers these individuals need. The fog up may be scaled to support a big range involving service requests. Mist up technology ultimately takes care of the micro-life cycle of applications and allows application administrators to focus on utility expansion and supervision. The Cloud Computing Program includes a new wide selection of products to create, check, run, work and shape applications within just typically the Cloud. Examples regarding fog up computing platforms encompass Amazon online Web Services, Google Software Engine and Microsoft's House windows Azure Platform.

The first Cloud processing type is infrastructure-as-a-service (IaaS), which is used with regard to Internet-based access to safe-keeping and computing power. Typically, the most basic category involving Cloud computing types, IaaS permits you to rent IT system – servers and online machines, storage, networks and operating systems – coming from a Cloud provider on the pay-as-you-go basis.

The second Cloud computing type will be platform-as-a-service (PaaS) that provides developers the various tools to construct and host web programs. PaaS is designed to be able to give users access to be able to the components they need in order to quickly develop and work web or mobile programs over the Internet, without having worrying about setting upward or managing the real infrastructure of servers, storage area, networks and databases.

The next Cloud computing type is usually software-as-a-service (SaaS) which is usually used for web-based apps. SaaS can be a method regarding delivering software applications on the Internet where Cloud providers sponsor and manage the software program applications making it less difficult to have the identical application on all involving your devices at the same time by simply accessing it inside the fog up.

2.2.4 Big Data Analytics

Big Data technology is based on previous data analysis methods and employs statistical methodologies such as regression, principal component analysis and other similar techniques. It provides real-time analysis by combining high-speed data retrieval with sensor information. The subjects of information science, data science, statistics and mathematical models are all multidisciplinary fields that may be studied together or separately. Statistical methods such as regression, multivariate statistical and multidimensional statistical, as well as mathematical knowledge and experience, are used to construct equations. The test on bog data revealed that increased technology has resulted in a massive amount of Big Data, and that academics are in the process of finding new paths of evaluating this data in order to extract valuable information. Investigators and users are no longer interested in learning what occurred and why it occurred (descriptive analytics), but rather in learning the answers to questions like what is occurring now and what is anticipated to occur in the future (predictive analysis). As a result, business analysis can be separated into three areas, i.e. descriptive, prediction and prescriptive insights, as indicated in Figure 2.3. In the next section, we will build predictive analysis, taking into consideration the significance of this technique for many stakeholders in society and business. The significance of Big Data analytics in smart city applications are discussed in paper [15–17].

Data is generated in a matter of hours with terabytes and petabytes of storage. Traditional database management systems can't keep up with the needs of storing, analyzing and administering vast amounts of data from an extensive series of sources. Figure 2.4 shows

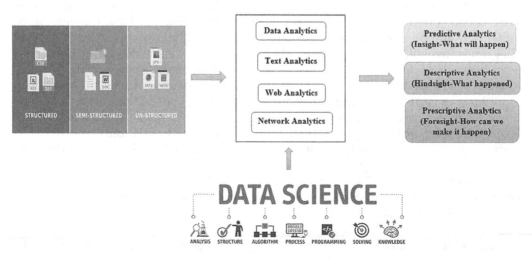

FIGURE 2.3
An existing study of predictive analytics.

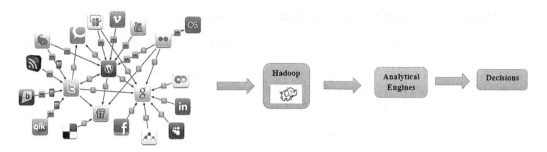

FIGURE 2.4
Architecture of Big Data analytics.

the simplified architecture of Big Data and analytics and shows how enormous dimensions of information are engendered from many bases, as well as how insights may be used to help in decision-making.

Predictive analytics uses data mining, pattern identification, statistics, deep learning, machine intelligence and information gathering to find relevant patterns in data. It also denotes the straightforward application of data analysis tools to solve issues or answer questions. Cutting-edge analytical approaches are used. Business intelligence (BI) and Big Data, when integrated with statistical methodologies, are making tremendous progress. Business intelligence methods aid in the study of data from multiple sources, allowing managers to make well-informed decisions. Experts in the field of research design the questions and variables; however, in predictive analysis, the model and association decisions are made based on the data. It's a method of systematic analysis in which a computer program searches for patterns and inherent relationships among variables of the study in a dataset. It is proposed to select the optimal connection regression coefficients for each link in order to reduce model errors. The technique makes considerable use of complicated information systems and requires numerous iterative cycles for best results.

Text messages, music, pictures and videos, among other things, make up the bulk of social media data. Text, videos, pictures, music and other forms of communication are used to share information. The typical process that consumers undertake while purchasing items is shown in Figure 2.5. They also share their thoughts and social media posts on the product, both before and after it has been bought. Companies may use technology to help them achieve their business objectives and make tactical decisions in support of those objectives. The processing ability based on predictive study is depicted in Figure 2.5. Next to that, we will discuss various ways in which social media analysis may be beneficial to businesses in the following part.

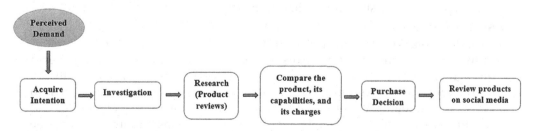

FIGURE 2.5
Processing based on predictive study.

2.2.5 Opportunities and Difficulties in Big Data Analytics

Our goal is to establish an open source Big Data enabling technology solution architecture, and the following are the primary principles for developing a strong Big Data Analytics platform. Data Archiving: A centralized solution for analytic archiving that allows data to be kept online for analytics rather than being moved offline, as well as seamless data life-cycle management and cataloging of information assets.

2.3 Efficacy in Relation to Time

Individuals who are able to extract value from a particular data source in real time may benefit from data from organized or unstructured sources. There are many examples of real-time data being used to address problems. Natural disasters may disrupt vital supply networks. In nations like India, where the logistics system is fragmented, the infrastructure is weak, and many large and small companies are engaged, the impact is larger. There is a need to proactively monitor supply chain risks in real time in order to avoid, react and prevent future disruptions and occurrences.

1. **Difficulties with Huge amount of Data**

   Because of its enormous scale and scope,Big Data presents some particularly difficult problems. The vast majority of companies are overwhelmed by the amount of data being generated at such a fast pace. The toughest challenges are measurement inaccuracies, as well as size and quality issues. The tremendous volume of data collected and stored every day at lower costs drives Big Data. It is an efficient statistical method. Noise, heterogeneity, correlation and inefficiency are all issues associated with Big Data that must be addressed.

2. **Data Size and Quality**
   Due to the expansion of the data, handling large amounts of it is problematic. Traditional databases can be managed with Excel or even other ETL software, but Big Data necessitates the use of a specialized technology framework such as Hadoop.

3. **Bigger Space**
   With so many variables, it's easy to create a misleading correlation, allowing some variables to mislead you about their relationship to the model outcome. Nonessential variables are likely included in the model. Large sample sizes result in high processing costs and algorithmic instability.

4. **System Reliability**
   Data dependability is another element to consider. Unstructured data is notoriously untrustworthy, prone to errors and data loss. The information is gathered through social media, smartphones, emails and text messages. Data is gathered from a variety of sources, including computer systems and industrial sensors. To extract tiny amounts of valuable information, data mining takes a long time and requires a lot of unlinked data. As a result, it's really taxing. Hunting for a needle in the haystack could be difficult.

5. **Data precision and adequacy**

   Although the dataset is vast and multidimensional, it does not reflect the entire population. When using social media as a basis of massive, unstructured data, only those who connect or participate in the online conversation are included in the data. Similarly, platforms such as Twitter restrict access to restricted datasets based on pre-defined requirements. Any biases in data interpretation should be controlled by researchers.

6. **Analytics Setup**

   Like any other IT project, Big Data Analytics has its own set of problems that must be handled inside an organization. With analytics come increased starting expenses, changing business trends and qualified information specialists. Implementing Big Data Analytics may be difficult due to many factors like data quantity and completeness issues, a lack of business support, insufficient personnel and knowledge and technological issues in the database, to name a few. There are a variety of additional problems associated with Big Data, including data cycle management, data redundancy, analysis, data secrecy, energy management, collaboration and data confidentiality. The fact that all of the aforementioned inquiries need a predicate that is well known results in a threefold increase in the number of workers who use the predicate in question. It is necessary to calculate a predicate-base on the resultant predicate index.

2.4 Why Cloud of Things

The particular numbers of connected equipment have recently exceeded the complete population from the Earth, and this is expected to increase even faster. With Net 3, the web is definitely reaching almost everywhere, and even the number of linked devices is increasing some sort of rise in the files generated. With IoT stuttering, it certainly plays a role in some sort of big chunk of Major Data [18]. Environmental sensors, supervising sensors, various actuators just about all generate data at amount, variation and speed. This is not possible in order to process data in past due IoT. This is in which Cloud computing comes within [19, 20]. Both IoT and fog up computing have seen distinct developments. But their incorporation has its own discussed interests that are known in the literature and can be noticed in the particular years to come. This seeks the concept involving the Assimilation Cloud regarding Things (COT) or Cloud-IoT example.

2.5 Integration Involving Cloud and Internet Associated with Things

Integrated IoT and Cloud computing produce a brand-new model, which we title cloud-IoT here. The specific worlds of Cloud and even "IoT" have seen neutral evolution. However, many mutually beneficial results obtained by their aggregate have already been identified in the literary works and might be resolved in the future. Particularly, IoT can easily benefit from definitely limitless skills and resources to be able to overcome the technical hurdles of the Cloud (e.g., storage, processing and power). Essentially, the fog up presents

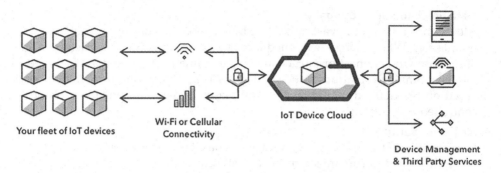

FIGURE 2.6
The application situations driven by the Cloud-IoT paradigm.

a green reply for packages that work with IoT service control and even composition and the records or even records they create. On the other hand, fog up can benefit from IoT to deliver more products in more dynamic and dispersed ways in a variety of real-world scenarios. [21, 22]. These kinds of worlds are really unique by each other as well as much better, their traits are often contributory. Such complementarity is an essential cause for most researchers in order to point their integration, to be able to enjoy the particular software program conditions.

Almost all the papers within just the literature do right now not sincerely see typically the Cloud in a larger context, meaning they believe in that the Cloud floods in some gaps within IoT (e.g., constrained storage). Alternately, search for IoT cloud bursting bursts (typical size). Due to the form of equipment, technology and techniques IoT has, it lacks the particular essentials, for example vulnerabilities, interoperability, flexibility, reliability, efficiency, supply and security. In simple fact, cloud-IoT enables the particular drift between facts selection and statistics processing, and even permits for quicker set up and integration of new content even as trying to keep expenses low for application and complicated information running. As a result, more complex analyses are possible, and information-driven decisions with testing algorithms can be used at a minimal cost to assist boost income and lower risk. A sample application driven by IoT paradigm is illustrated in Figure 2.6.

Cloud-IoT has provided new smart services and packages which have the sizeable impact on everyday lifestyles. Machine-to-Machine Communication (M2M) is beneficial for the majority of such applications outlined here, not just when items need to be updated and then sent to the fog up. These packages include the following:

a. Wellbeing care
b. Smart cities and even communities
c. Smart homes and smart metering
d. Video security
e. Motor and cognitive range of motion
f. Smart power and intelligent grid
g. Smart Strategies

2.5.1 A Couple of Cloud-IoT Applications

The Cloud-IoT approach has got released several programs and intelligent products and services which have impacted end customers' daily existence [5]. The following offers a brief conversation of certain programs which were improved from the Cloud-based IoT paradigm.

a. Healthcare

   This has resulted in huge number of applications in medical domain.
b. Smart Places

   The middleware of sensible cities for future years can be offered through IoT, retrieving files from facility acquisitions, IoT systems and information transparency.
c. Smart Houses

   A lot of iCloud-enabled IoT software have empowered automation for house tasks, where in fact the acquisition of varied embedded gadgets and Cloud processing enabled internal operation (e.g. security management, good metering, energy preservation).
d. Video monitoring

   By implementing Cloud-based IoT, wise training video monitoring will be in a position to manage, retail and process video tutorial articles from video tutorial receptors effectively and effortlessly; which will automatically obtain the details from the scenes.
e. Smart and Automotive Mobility

   The integration of Cloud processing into the Global Positioning Method assists in smart traffic monitoring and avoiding congestion.
f. Smart vitality and wise grid

   Cloud-IoT offers a variety of solutions in renewable energy sector.
g. Smart logistics

   It permits and reduces the programmed activity of products between customers and manufacturers, while at the same time facilitating the traffic monitoring of freight (e.g., the shipment industry, traffic monitoring).

2.6 The "QoS" on Integration of Cloud and IoT

Heterogeneous networks are (by approach to default) a versatile valuable resource to offer several support or service [23, 24]. This does not indicate that there aren't still a variety of visitors to the community, nor does it disprove the ability of a single community to accommodate all applications without sacrificing the quality of the Service (QoS). There are two instructions of application: Navigation within and reducing the amount of visitors delays (e.g. for checking reduced sample rates), and bandwidth and crucial delays inelastic (real-time) traffic (e.g. noise or website visitors monitoring), which may be inside addition discriminated against specifics associated services (e.g. Gary the gadget guy, high-resolution, low-decision videos) along with specific QoS needs. For that reason, a controlled, appropriate approach to offer unique community vendors, each as well as the very own quality involving service necessities is expected. It isn't always soft to be able to offer QoS authentication upon Wi-Fi networks, as components often cause "gaps" throughout device authentication because relating to useful resource allocation and even management troubles on distributed Wi-Fi media.

Service quality within Cloud computer is any kind associated with other major region related with studies in order to need more interest as specifics in addition to equipment that are essential within the particular Cloud. Dynamic making strategies and useful resource share algorithms primarily based on the particle production method are being produced along with high electricity programs, so when IoT grows, this special could become a bottle of wine. However, concurrently as jogging heterogeneous networks are (by using default) an adjustable aid; to provide more compared to one services or supplier is not to admit that the most successful types of visitors are no longer present in the community, and also that the community as a whole is able to support all applications without sacrificing the quality of the Support (QoS). You can easily find two instructions involving application: navigation and minimization of traffic delays (e.g, for checking reduced sample rates), along with bandwidth and critical delays in elastic (real-time) site visitors (e.g. noise or possibly traffic monitoring), which could be more discriminated towards statistics connected services (e.g., high-decision low-decision videos) with special QoS essentials. Consequently, a controlled, correct solution to offer unique network vendors, each having its very own QoS needs should be used. It isn't constantly clean to supply QoS authentication on wireless networks, considering that components frequently cause "gaps" in device authentication since a result of beneficial resource allocation and management troubles on contributed Wi-Fi media. Service High top quality in Cloud computing will be usually every other key place regarding research in order to require even more consideration as records plus resources are to be experienced inside the Cloud. Way producing plans and support share algorithms based completely on the particle production method are developed. Collectively with high electricity deals so that as IoT grows, this can switch into a bottle. Nevertheless, as working with typically the Clouds has excellent advantages; at the same time, you can find problems in the region involving provider quality (QoS). Quality of service relates to the sums of efficiency, reliability and availability supplied while using working device or system or even hosted facilities.

QoS is crucial for Cloud customers who count on businesses to be able to deliver on their promises of sophisticated features as well as for cCoud service providers who are trying to achieve the ideal balance between QoS standards and operating costs. Any breach within the assistance degree arrangement (SLA) involves a reduction to each Cloud firm and fog up users. Surgery is frequently accepted with the help of firms as a method related to pleasing the SLA, on the other hand it fails to maximize using resources, especially the exclusive Cloud.

Existing methods for the provision and help of service level deals in IoT in improvement to Cloud are extremely restricted. Within the context of the Cloud IoT, these types of processes will need to be reconsidered or remodeled to meet the issues posed by future Cloud-IoT applications. On the other hand, QIS Cloud-IoT qualification is usually expected to be challenging, together with emerging discipline. This is definitely usually due to typically the absence of standard, end-to-end quality of service authentication methods (between end-user, IoT and Cloud devices), typically the complexity of multi-layer in use, as well as the presence regarding the plethora of QoS concerns and parameters in every single layer. From the perceptive of Cloud-IoT software, we predict that the traditional QoS authentication technique may not be sufficient, network communication is crucial, but it is a compact thing. Cloud-related parameters (I/O efficiency and COMPUTER usage rate), devices (battery), network sort and program will have to become used inside conjunction together with network variables (bandwidth, postpone and jitter) to appear for the overall Quality of service associated with applications intended for Cloud-IoT. Obtaining a good view of Cloud-IoT, specially the sensing paradigm, is usually also essential to realize and carefully develop quality of service metrics and associated SLAs that take into bank account the complexity presented by just each layer.

2.7 The Benefits of Converging IoT-Cloud

Considering that the IoT suffers from nominal capability with regard to operational capability and safe-keeping, it should also manage issues with total performance, privacy, privateers and consistency. IoT assimilation throughout the Cloud is typically the acceptable manner to overcome a large number of barriers associated with IoT. Cloud can in fact gain IoT by technique of increasing its prospect with real-world packages inside a powerful and dispersed approach and offering brand fresh offerings for tens with regard to countless gadgets in amazing health contexts. In improvement, iCloud offers easy-to-use and low-fee utility and end-user services. Cloud additionally tends to make simple the drift and the collection of IoT stats and provides speedy, less expensive set up and the usage of intricate information control and distribution. The advantages of adding IoT in the Cloud are normally discussed with this area next.

1. Communication

   Two crucial components of the Cloud-based IoT paradigm are app sharing and information. The Internet of Things (IoT) has the potential to transmit prevalent services, whereas software is frequently employed for distribution and small-scale data collection. Through the use of integrated services and systems, fog up is a reliable and economical solution that can be used to connect to, manage, and play anything. [25]. Availability associated with speedy systems enables effective tracking and management associated with far-flung objects, in addition to gaining access to real details. That is really well worthwhile citing that, although Haze up can appreciably help make larger and leverage IoT online connectivity, it's still much weak in a few places.

2. Storage

   Given the likelihood of IoT to be utilised on billions of devices, this has a vast array of data extracts, resulting in enormous amounts of arbitrary or random information. When it comes to handling the significant amount of information produced by the IoT, the Cloud is thought to be one of the most practical and affordable alternatives. It also creates new opportunities regarding data integration, the usage, and revealing with 0.33 functions.

3. Processing

   The strength involving IoT devices is characterized by simply means of lowered dimension abilities that block typically the area and general overall performance of complex data. Alternatively, the information accumulated will be utilized in high-professional locations; certainly this particular chapter is where consolidation and dissemination are done. However, accomplishing scalability remains a new quest without proper structure. To realize an option, Cloud gives unlimited electronic digital working talents as nicely as the desired running model. Predictive algorithms and information-driven decision-making can become included into IoT in order to maximize product sales and minimize risks from a new low worth.

4. Scale

   Together with billions of customers attaining every other and different statistics being gathered, the particular sector is quickly relocating to the Internet of Everything (IoE) area directs a network of systems and physical objects that generate new opportunities and potential threats. A fully Cloud-based, fully IoT approach offers fresh concepts and products based on expanding the Cloud with IoT features. , which usually allow Cloud to job with real-global new conditions and guide to typically the emergence associated with modern services. It offers a variety of features, including ease of use, seamless access, inexpensive delivery and delivery rates, and ease of use.

New versions:

The use of Cloud-based IoT enables new contexts regarding clever gadgets, programs and offerings. A few brand-new models are categorized as following:

i. SAAS (Sensitivity love a service), which permits entry to sensor data;

ii. EAAS (Ethernet like a Service), the primary function concerning offering broadband connectivity so as to govern far-flung gizmos;

iii. S-AaaS (Sensitivity and Kinesis as a Service), which often gives automation control abilities;

iv. IP-MaaS (identification and insurance plan control being a service), It frequently gives access to the security and administration of identification documents;D-BaaS (Database since a Service), which gives intelligent statistics management;

v. S-EaaS (Sensor Event as the Service), which sends messages to providers generated with typically the support of sensor situations;

vi. Se-naaS (Sensation being a Function), which provides faraway realizing control;

2.8 One Cloud-Based IoT Architecture

Relative to many preceding studies, this construction of recognized IoT gizmos is typically split directly into three one-of-a-kind tiers: energy, know-how and system. Many experts claim that the area layer is the errors up layer in the Cloud-based IoT architecture. [26]. The cognitive coating is usually utilized to discover devices and collect data, gathered from its area. In contrast, the important purpose linked to the particular network would be able to transfer the accrued information to the Internet as well as iCloud. Finally, the electricity layer provides loads associated with solution connectors.

2.9 Difficulties in Converging IoT-Cloud

a. Secrecy:

Moreover, the different difficult situations require a few attentions; for example, a distributed gadget is characterized by means of some of potential attacks, along with SQL injection, consultation rides, cross-bite writing and the closest channel. In addition, high-risk, which incorporates consultation hijacking and virtual device break out is likewise problematic.

b. Heterogeneity

This is one of the most important challenges going through the Cloud-based IoT technique issues. In addition, the challenge of heterogeneity can increase when quit users adopt multiple Cloud solutions, and for this reason services will rely on more than one provider to enhance utility performance and robustness.

c. Big records

There are a variety of predictions that Big Data drive holds 50 billion IoT gadgets by means of 2020, it's essential to pay close interest to travel, access, garage and processing at the expense of big records with a view to be produced. Indeed, if we're technologically advanced, it's far clear that IoT will be one of the principal assets of large data, and that iCloud is in a position to carry out long-term information storage in addition to incorporating it into complicated analysis. The managing of large quantities of records generated is a major problem, for the reason that operation of the utility all relies upon on the houses of this records control service. Finding the right records management solution with the intention to permit Cloud to handle big quantities of statistics remains a chief problem. In addition, information integrity is an essential factor, not only due to its effect on provider quality but also because of safety and privacy problems, a lot of which relate to the statistics submitted.

d. Performance

As a result, obtaining sufficient community performance to transfer records to Cloud environments is a huge challenge; indeed, this is due to the fact that the expansion of broadband is incompatible with long-term sustainability and environmental protection integration. In many circumstances, services and information must be available in a timely manner. This is because time travel can be excruciatingly painful.

Problems and real-time programmers have a huge impact on productivity. Transferring large quantities of statistics generated from IoT gadgets to the Cloud requires a large bandwidth. Because of this, the major difficulty is getting sufficient community performance to transfer records to Cloud environments; indeed, this is due to the fact that Broadband's growth is inconsistent with sustainability and ecological integration. In many cases, services and information delivery is required in high overall performance. This is because time travel can sufferng from unbearable troubles and real-time programs significantly affect efficiency.

e. Legal elements

Legal aspects are very crucial in recent studies on precise programs. For example, carrier providers need to adapt to distinct worldwide laws.

f. Monitoring

Monitoring Cloud Computing's first action while it comes to performance, useful resource management, potential planning, despite the fact that there are still some related challenges associated with velocity, volume and diverse factors of IoT.

g. Big scale

As the enormity of emerging packages increases, there arise a slew of new problems to contend with. For example, reaching consolidation electricity and storage capacity requirements will become hard.

2.10 Conclusion

In summary, this chapter provides a brief introduction about what the IoT is and what its public relations are and how this has turned out to be a significant and indispensable thing in life. It also presents the history of the rise of the IoT from the history of its origins to this day, and then provides a simple summary of what Cloud computing is, what does it do and how does it work. These are the key points and a simple summary is illustrated in the figures on the structure and organization of Cloud computing and also its types. The chapter presents the integration between the Cloud and the Internet related to things and the extent of the connection between them with Big Data-based Solution. Then we present the most important considerations for the applications linked among Cloud-IoT. Applications are also very important. We provide quality service as well as the link between them, and we make a simple comparison showing the most important points and the distinction between the Internet stuff and Cloud computing. Then we provide the basic benefits of the incorporation of the Internet stuff and Cloud computing, and we also provide important challenges they face.

References

1. Alamri, A., Ansari, W. S., Hassan, M. M., Hossain, M. S., Alelaiwi, A., Hossain, M. A., 2013. A survey on sensor-cloud: architecture, applications, and approaches. *International Journal of Distributed Sensor Networks*, 9(2), 917923.
2. Alhakbani, N., Hassan, M. M., Hossain, M. A., Alnuem, M., 2014. A framework of adaptive interaction support in cloud-based Internet of Things (IoT) environment. In: *Internet and Distributed Computing Systems*. Springer, pp. 136–146.
3. Antonic, A., Roankovic, K., Marjanovic, M., Pripuic, K., Zarko, I., Aug 2014. A mobile crowd-sensing ecosystem enabled by a cloud-based publish/subscribe middleware. In: *Future Internet of Things and Cloud (FiCloud), 2014 International Conference on*. pp. 107–114.
4. Armbrust, M., Fox, A., Griffith, R., Joseph, A. D., Katz, R., Konwinski, A., Lee, G., Patterson, D., Rabkin, A., Stoica, I., et al., 2010. A view of cloud computing. *Communications of the ACM* 53(4), 50–58.
5. Atkins, C., et al., 2013. A Cloud service for end-user participation concerning the Internet of Things. In: *Signal-Image Technology & Internet-Based Systems (SITIS), 2013 International Conference on IEEE*, pp. 273–278.
6. Atzori, L., Iera, A., Morabito, G., 2010. The internet of things: A survey. *Computer Networks*, 54(15), 2787–2805.
7. Ballon, P., Glidden, J., Kranas, P., Menychtas, A., Ruston, S., Van Der Graaf, S., 2011. Is there a need for a cloud platform for european smart cities? In: *eChallenges e-2011 Conference Proceedings, IIMC International Information Management Corporation*, Italy.
8. Karthikeyyan, P., Velliangiri, S., 2019. Review of Blockchain based IoT application and its security issues. In: *2019 2nd International Conference on Intelligent Computing, Instrumentation and Control Technologies (ICICICT)*. Kannur, Kerala, India, vol. 1, pp. 6–11, IEEE.
9. Bernaschi, M., Cacace, F., Pescape, A., Za, S., 2005, February. Analysis and experimentation over heterogeneous wireless networks. In: *First International Conference on Testbeds and Research Infrastructures for the DEvelopment of NeTworks and COMmunities* (pp. 182–191). IEEE.
10. Bhattasali, T., Chaki, R., Chaki, N., 2013. Secure and trusted cloud of things. In: *India Conference (INDICON), 2013 Annual IEEE*. IEEE, pp. 1–6.
11. Biswas, J., Maniyeri, J., Gopalakrishnan, K., Shue, L., Eugene, P. J., Palit, H. N., Siang, F. Y., Seng, L. L., Xiaorong, L., 2010. Processing of wearable sensor data on the cloud—a step towards scaling of continuous monitoring of health and well-being. In: *Engineering in Medicine and Biology Society (EMBC), 2010 Annual International Conference of the IEEE*. IEEE, pp. 3860–3863.
12. Bitam, S., Mellouk, A., 2012. ITS-cloud: Cloud computing for Intelligent transportation system. In: *Global Communications Conference. (GLOBECOM), 2012 IEEE*. IEEE, pp. 2054–2059.

13. Bo, Y., Wang, H., 2011. The application of cloud computing and the internet of things in agriculture and forestry. In: *Service Sciences (IJCSS), 2011 International Joint Conference on*. IEEE, pp. 168–172.

14. Bonomi, F., Milito, R., Zhu, J., Addepalli, S., 2012. Fog computing and its role in the internet of things. In: *Proceedings of the First Edition of the MCC Workshop on Mobile Cloud Computing. MCC '12. ACM*, New York, NY, USA, pp. 13–16.

15. Botta, A., Pescape, A., Ventre, G., 2008. Quality of service statistics over heterogeneous networks: Analysis and applications. *European Journal of Operational Research*, 191(3), 1075–1088.

16. Carriots, 2014. https://www.carriots.com. Castro, M., Jara, A., Skramstad, A., March 2013. Smart lighting solutions for smart cities. In: *Advanced Information Networking and Applications Workshops (WAINA), 2013 27th International Conference on*. pp. 1374–1379, IEEE.

17. Aazam, M., Huh, E.-N., Aug 2014. Fog computing and smart gateway-based communication for cloud of things. In: *Future Internet of Things and Cloud (FiCloud), 2014 International Conference on*. pp. 464–470.

18. Abdelwahab, S., Hamdaoui, B., Guizani, M., Rayes, A., 2014. Enabling smart cloud services through remote sensing: An internet of everything enabler. *Internet of Things Journal, IEEE*, 1(3), 276–288.

19. Velliangiri, S., 2020. An enhanced multimedia video surveillance security using wavelet encryption framework. *Journal of Mobile Multimedia*, 15, 239–254.

20. Aitken, R., Chandra, V., Myers, J., Sandhu, B., Shifren, L., Yeric, G., 2014. Device and technology implications of the internet of things. In: *VLSI Technology (VLSI-Technology): Digest of Technical Papers, 2014 Symposium on*. pp. 1–4.

21. Akyildiz, I. F., Su, W., Sankarasubramaniam, Y., Cayirci, E., 2002. Wireless sensor networks: A survey. *Computer Networks*, 38(4), 393–422.

22. Sathiyaraj, R., Bharathi, A., Balamurugan, B. 2022. *Advanced Intelligent Predictive Models for Urban Transportation*. Chapman and Hall/CRC.

23. Sathiyaraj, R., Bharathi, A. 2020. An efficient intelligent traffic light control and deviation system for traffic congestion avoidance using multi-agent system. *Transport*, 35(3), 327–335.

24. Rajendran, S., Ayyasamy, B. 2020. Short-term traffic prediction model for urban transportation using structure pattern and regression: An Indian context. *SN Applied Sciences*, 2(7), 1–11.

25. Bharathi, A., Balamurugan, B., Chokkanathan, K., Sathiyaraj, R., Singh, A. 2019. Internet of things technologies. In Sudan Jha, Manju Khari, Raghvendra Kumar (Eds.), *Internet of Things in Biomedical Engineering* (pp. 291–322). Academic Press.

26. Muthuramalingam, S., Bharathi, A., Gayathri, N., Sathiyaraj, R., Balamurugan, B. 2019. IoT based intelligent transportation system (IoT-ITS) for global perspective: A case study. In Valentina E. Balas, Vijender Kumar Solanki, Raghvendra Kumar, Manju Khari (Eds.), *Internet of Things and Big Data Analytics for Smart Generation* (pp. 279–300). Springer, Cham.

3

Security and Privacy Challenges and Solutions in IoT Data Analytics

Kumar Shalender

Chitkara University, Rajpura, India

Rajesh Kumar Yadav

Amity University, Noida, India

CONTENTS

3.1 Introduction 43
3.2 Security Challenges with Introduction of IoT and Big Data 45
3.3 IoT Challenges in Smart City Applications and Primary Variables 46
    3.3.1 Storage and Management of Data 47
    3.3.2 Issues Related to Data Visualization 48
    3.3.3 Data Privacy and Confidentiality 48
    3.3.4 Data Integrity 49
    3.3.5 Security of the Devices 49
    3.3.6 Issue of Power Supply 50
3.4 Solutions Offered for IoT Security Systems for Smart City Applications....50
    3.4.1 Secure IoT System and Network 50
    3.4.2 Using Authentication for IoT Systems 51
    3.4.3 Using IoT Encryption Technology 51
    3.4.4 IoT Analytics for Security 51
    3.4.5 Testing Hardware 51
    3.4.6 Developing Secure IoT Apps 52
    3.4.7 Keeping Abreast with the Latest Security Threats 52
3.5 Conceptual Framework for IoT Data Analytics 52
3.6 Conclusion 54
References 54

3.1 Introduction

The concept of the Internet of Things (IoT) has been around for a couple of years and increasingly becoming prominent with the advancement of technology and the widening net of internet availability. IoT can be considered as a system of interconnected devices and objects which communicate with the help of internet connectivity. The devices can be wired or function in a completely wireless manner depending upon the product and

DOI: 10.1201/9781003217404-3

process architecture they are based upon. The reason behind their increasing popularity is not difficult to ascertain. These devices and systems are used in a number of applications used in almost every organization today, and the application scope of these is spread across industries (Khvoynitskaya 2020). Be it the telecommunication industry, FMCG sector, or retail industry, every stakeholder in the ecosystem of these business segments requires one or other kind of connected devices and systems to perform with the desired efficiency and effectiveness. The same is true for the manufacturing and supply chain and logistics industries where IoT systems are specifically beneficial for making sure that goods and services are produced and delivered on time and the overall system can deliver higher operational efficiency and functional effectiveness (Monther and Tawalbeh 2020). Even in the automobile industry, right from the logistics to supply chain and from the engine management to body architecture, IoT systems are ruling the roost (see Figure 3.1). Therefore, it comes as no surprise that IoT devices command enormous importance and significance among the stakeholders spreading across the industries.

The term IoT was introduced first by Kevin Ashton when the author was working on the radio frequency identification concept (RFID) back in the year 1999. An RFID system consists of a number of sensors and actuators which were embedded in the system to transmit the data although the original idea behind the interconnected devices was sprung in the 1960s. At that point in time, this concept was known by the name of embedded internet (aka pervasive computing). Ashton talked about the concept of IoT in the context of enhancing the efficiency of the supply chain, but it is only in the decade of 2010 that the concept of IoT received wide popularity in a true commercial manner (Meng et al. 2018). Among the policymakers, the Chinese government was the first one that attached highest importance to the concept of IoT by announcing a budget running in billions of dollars to promote the concept of interconnected devices. After embracement from the Chinese government, regulators and policymakers across the globe started putting emphasis on IoT that led to a revolution in the popularity of the concept and its practical applications. The applications of IoT have brought a sea change in enhancing the efficiency and effectiveness of industries across the spectrum. On the individual level, IoT has improved the lifestyle of

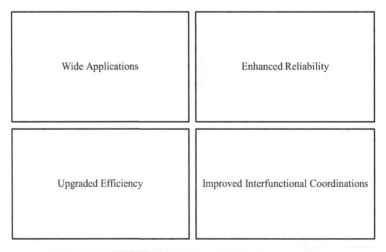

FIGURE 3.1
Uses and benefits of IoT devices: Author's own conceptualization.

consumers by offering numerous ways to automate things while on the broader scale of industrial application, it has changed the face of companies by offering them effective and efficient methods of conducting their business processes (Hassan 2019).

3.2 Security Challenges with Introduction of IoT and Big Data

However, the enhanced application of the IoT concept has also brought to light a number of challenges related to the data privacy and security of the users. As these devices keep on collecting information on various parameters, including the sensitive ones pertaining to personal and professional lives of consumers, these pose challenge for both organizations and users on how to keep their data safe and out of reach of the hackers (Ali et al. 2018). These challenges, which are addressed under the theme of cybersecurity risks, are very critical and if not addressed properly can easily result in huge financial as well as non-monetary losses for the stakeholders who are part of the ecosystem. Using these devices without exercising discretion, not updating the devices to the latest safety norms, using the old passwords, not changing the passwords at regular intervals, and accessing dubious websites or applications through these devices can result in an enhanced cybersecurity risk for the stakeholders involved in the ecosystem of IoT devices (Liu et al. 2017).

Not exercising caution while using IoT devices can lead to a data breach and most of the experts agree that without conscious behavior from users, IoT devices are vulnerable to data theft and attacks. It is also unfortunate that despite having numerous safety protocols that can be applied to effectively reduce the vulnerability of IoT devices, most of the procedures have not been documented which, in turn, restrict their applicability and capacity to safeguard the data (Tawalbeh and Tawalbeh 2017). This also restricts the access of these mechanisms to users, making them easy targets for hackers to exploit their sensitive data and use the information for carrying out financial crimes.

Since the 2008 financial crisis, hackers keep on developing and updating malware to infect the IoT devices of users and while device manufacturers try hard to avoid these attacks, hackers seem to be one step ahead with their advanced technology and capabilities to exploit the vulnerabilities in the system. The threat of data breaches has become even more prominent and a cause of concern when it comes to industrial applications of IoT devices. As the adoption of these interconnected devices keeps expanding on a wider scale, stakeholders in these companies are increasingly concerned of the safety and security of their interconnected systems. Any kind of breach in the IoT system can have a significant impact on the efficiency and procedural mechanism of the organization, and hence these companies spend a lot of money to safeguard their IoT system with the latest state-of-the-art technology and regular updating techniques (see Figure 3.2). In the process of safeguarding the IoT devices, organizations hire and deploy professionals who can provide them protection against these threats and develop a comprehensive safety mechanism within the company to prevent attacks on their systems.

Take, for instance, the concept of a Smart Kitchen enabled by IoT devices (Bugeja et al. 2016). Hackers can easily exploit the vulnerabilities of the system and can get access to the sensitive information in the absence of a credible safety mechanism. More concerning issue is the upcoming 5G technology in India as it's going to change the way communication between interconnected systems happens today. The speed is going to be enormous when

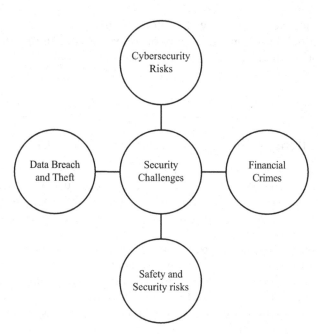

FIGURE 3.2
Security challenges with IoT and big data: Author's own conceptualization.

we talk about 5G capabilities, which means that threats to data security and privacy will be even more pressing than the concerns we have related to IoT devices today.

In the context of this background, this chapter deals with the primary issues related to security and privacy challenges in IoT Data Analytics and then suggest solutions to how to overcome these challenges in the context of changing business environment and ever-upgrading technology. The rest of the chapter is divided into the following sections: Section 3.3 talks about the challenges related to data safety and security of IoT devices and identifies important variables in IoT systems that require specific attention to fight data breaches and hacking attempts. This is followed by suggested solutions that can help organizations to effectively preserve the safety and security and guide the industry to come up with IoT systems with enhanced safety and security features.

3.3 IoT Challenges in Smart City Applications and Primary Variables

It is very much obvious that with thriving IoT ecosystem, the challenges related to these interconnected devices in the context of smart city applications are also on the rise. There are numerous issues that need to be addressed including the ones related to data security, privacy, storage and management, data integrity, and the overall security of the IoT devices. In addition, there are other challenges that organizations involved in the Data Analytics of IoT devices have to contend with while exploring the application potential of these devices in building smart cities of present and future. Significant among these are interoperability, dependability, flexibility, and adaptability of the devices on a global scale (Gemalto 2020). Amidst all these challenges, the following are the primary factors that

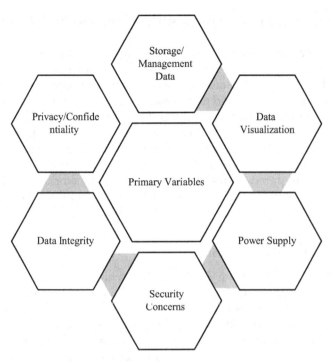

FIGURE 3.3
IoT challenges in smart city applications: Author's own conceptualization.

need to be considered in detail in order to address successfully the issues related to IoT devices, their connectivity, and storage capacity.

IoT systems keep on collecting the data from the interconnected devices which poses a huge challenge for the storage capacity and electronic waste management. The capacities of the IT infrastructure are also coming under a constant pressure from the enormous amount of data that is transferred between interconnected devices on a daily basis. The real-time analysis has also significant implications for many industries, especially the ones which are dependent on the real-time analysis of the data for their performance aspect. Following are the key challenges faced by IoT systems in their integration and application in the context of smart city applications that need to be addressed on urgent basis for a sustainable IoT system and its interconnected devices (see Figure 3.3):

3.3.1 Storage and Management of Data

It can be easily concluded that incessant amount of data that is captured, collected, and analyzed by the IoT system is overwhelming. The existing infrastructure and the storage mechanism have their own capacities and limitations and in the wake of more and more data being collected, the storage and management of data is increasingly becoming a difficult task to handle for data analytics companies involved in the collection and analysis of data captured by the IoT devices (Estrada et al. 2020). In fact, it has also emerged as a security challenge for the companies as data related to sensitive sectors such as transportation, supply chain, logistics, and telecommunication is very important and crucial in its application. There are many processes which work on inputs provided by IoT devices on a real-time basis and in case any kind of a discrepancy in the storage and management of the data, the working of the entire industry gets impacted in a significantly negative manner.

Therefore, it is very much essential for the companies involved in storage and management of the data to keep on enhancing their capacities on a regular basis so that the issue of the storage and proper management of the data can be handled in a desirable manner. In addition to continuous storage and management of data, the challenge related to dumping of the expired data is also among the significant issues faced by the industry today. Environmental activists are increasing the pointing towards the electronic waste management and without proper disposal policy, the IoT devices and their connected ecosystem will also face the problem related to the ecosystem degradation in the coming times. It is, therefore, very much required on the part of organizations working in data analytics to properly search for a solution that will help them not only with the storage and management of data but also with proper disposal of the expired data.

3.3.2 Issues Related to Data Visualization

This problem is specific to the Data Analytics firm which uses the data collected by the system of IoT devices for enhancing the operational efficiency of the companies across industries. The primary challenge here is that the data collected by IoT system is primarily unstructured or semi-structured in nature. Most of the time data is completely heterogeneous and it will require a range of processes that will make it more refined so that it can be used for the industry purposes. The data refinement is not an easy process and requires a comprehensive input on the part of the organization and requires a variety of financial and non-financial resources. Data analytics firms have to employ competent manpower in order to sift through the data and uncover the insightful information behind those numbers. This, in turn, has financial implications for the organization as recruiting and selecting competent employees for the specific purpose entail organization to spend its financial resources to hire these professionals and keep them with the organizations. Even more prominent challenge is to find the suitable manpower as demand in this field exceeds the supply (Tawalbeh et al. 2020). This puts potential candidates at a position of strength as they can leverage their position and bargain hard with the organization in terms of their salary and compensation. Therefore, this added cost comes as an additional challenge for the data analytics firms which are looking to enhance the operational efficiency and functional procedures of the companies with the help of analysis of the data collected by the ecosystem of IoT devices.

3.3.3 Data Privacy and Confidentiality

This is one of the primary challenges related to IoT ecosystem not only to the industry but also to the individual user who are using interconnected devices in their personal life. The constant process of collecting and generating the data by the ecosystem of IoT devices makes them vulnerable to the security issues. This is specifically true for the IoT devices which are used at home without any kind of a strict password protection and security compliance. Cybersecurity risks related to these devices are high and thanks to the advanced technologies and malware used by the hackers, it's not hard to compromise the security of IoT systems.

Even in case of big organizations, every now and then the industry witnesses some kind of a breach in the data privacy. Maintaining the confidentiality of the information is also very important aspect of the agreement between an organization and its clients. This is even more relevant when we talk about the information that is being constantly collected and generated by system of IoT devices. Although there is a confidentiality clause that is being signed by both client and the service provider, but still there are many instances when this confidentiality clause is breeched by one or the other party – presenting a signifi-cant security challenge for the data of the concerned stakeholders (Singh et al. 2016). It is, therefore, necessary to address the challenges related to the data privacy and confidential-ity because without properly addressing this challenge, the desired rate of growth in the sector cannot be achieved in the long term.

3.3.4 Data Integrity

Closely related issue to the data privacy and confidentiality is the integrity of the data that has to be maintained by all the stakeholders involved in the ecosystem of IoT devices. In fact, maintaining the integrity is one of the prerequisites for the growth of the sector and any kind of a breach or undermining of the integrity can have a devastating impact on the entire ecosystem of the industry. There are many tools, techniques, procedures, com-mands, and standard operating procedures that are being employed by the industry and their use have actually showcased their effectiveness in tackling the problem related to data integrity in the long run. In order to maintain their effectiveness, it is required that safety protocols should be upgraded on a constant basis so that any kind of data breach can be encountered with the properly laid down processes and procedures. The issue of data integrity is of paramount importance when it comes to the reputation of the data analytics firms. Any breach in the integrity of the data collected by the company can lead to long-term consequence for the organization. Hence, it is very much desirable in the part of data Analytics firms that they put into effect proper policy guidelines and communicate it throughout the organization so as to sensitize their employees and especially the people working in the domain of collecting, collating, and analyzing data of clients for developing and maintaining their reputation in the industry (Liyanage et al. 2020).

3.3.5 Security of the Devices

IoT devices are vulnerable to the security attacks, and therefore they require a compre-hensive safety mechanism in order to avoid any kind of a threat from the hackers. With the passage of time, we are now witnessing a credible number of safety solutions that can help protect the IoT devices, still here are many areas where the safety devices are vulner-able and can be easily exploited by the hackers. This is because users in the IoT ecosystem including organizations are not very serious about protecting these devices from the data hackers. Although the inbuilt safety mechanisms are good enough to protect these eco-systems, the lack of the awareness on the part of a user is the primary reason behind their compromise and exploitation by the hackers. The rules to protect IoT devices are not very complex and if adhered to by stakeholders, the security attacks on these devices can be reduced comprehensively. It is also required that all these procedures must be documented in the written form and customer should get information about these safety protocols so that implementation of these procedures can happen in reality (Ding et al. 2016). It is also required to constantly upgrade the devices to the latest software version as devices with the older safety protocols are more likely to get attacked by the malicious software or malware.

3.3.6 Issue of Power Supply

A less discussed, although equally important, issue is related to the power supply as IoT devices and their ecosystem require an interrupted electricity for their effective and seamless operations. This means these devices have their environmental footprints and if electricity used in working operations of IoT devices is generated from the non-renewable sources, it will have serious consequences for the climate change. The larger issue here is to decouple the IoT ecosystem from carbonized energy and then the whole ecosystem should become part of the decarbonized electricity generation (Izzat et al. 2020). The particular problem is going to have serious implications in the future as people are waking to the reality of climate change and governments too have started doing their bit in order to reverse the harmful impact of the processes and procedures adopted by them in the past.

3.4 Solutions Offered for IoT Security Systems for Smart City Applications

There are a variety of solutions that can be used to secure IoT security systems and in the wake of rising cybersecurity threats, these solutions must be integrated so as to make sure that safety and security of the IT system of the organization remain intact throughout the lifecycle of the organization. Following are solutions that can help the organizations cutting across the industry to achieve desirable levels of safety for IoT security systems (see Figure 3.4).

3.4.1 Secure IoT System and Network

At the very basic level, data analytics firms most secure their IoT systems and networks with the help of up-to-date antivirus, firewalls, anti-malware, and prevention and detection of intrusion systems. This seems to be the very basic and fundamental step but in reality, it can prove instrumental in providing enhanced protection to IoT structure and system

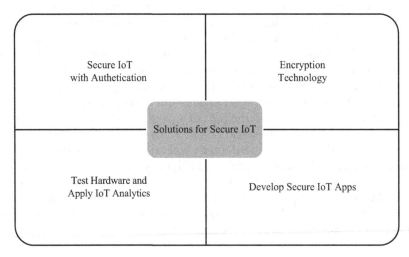

FIGURE 3.4
IoT security solutions for smart city applications: Author's own conceptualization.

within the organization. This first-layer of security discourages the intruders in making advances in the IoT system of the organization, and although this is no guarantee of immunity against attacks, these certainly act as a frontline protection against advances made by data hackers (Alaba et al. 2017). It is important to note that the protection offered by these basic security systems might not prove enough but in the long run, these are essential to strengthen the overall security credentials of the IoT network.

3.4.2 Using Authentication for IoT Systems

Using a multi-step authentication process for IoT systems is also a credible way to safe and secure IoT infrastructure in the organization. There are a number of measures that an organization can employ in order to implement the multi-step authentication procedure for securing access to IoT infrastructure. Some of the prominent examples include the use of biometrics, two-factor authentication, multiple-user management, and digital certificate among others (Khan and Salah 2018). The use of these robust authentication mechanisms will add to the safety and security of your IoT infrastructure and will help organizations to keep themselves safe from cyberattacks.

3.4.3 Using IoT Encryption Technology

Moving up the ladder of sophistication, Data Analytics firms must use encryption technology to transmit data between different IoT devices. By making use of the cryptographic algorithms, the transition of the data between different devices can be safely encrypted and kept away from the reach of hackers. Organizations can subscribe to full lifecycle management of the processes in order to enhance data privacy and in the process, making the IoT infrastructure completely safe and secure.

3.4.4 IoT Analytics for Security

The special feature associated with the analytics security of IoT devices will help your organization to keep the data safe and beyond the reach of hackers. The application of this feature will uncover the security patterns that might be detrimental to the security of IoT devices and hence place the firm in an advantageous position when it comes to securing the IoT infrastructure. The analysis report of the security systems can provide insightful information about the possible threats that can impact the IoT infrastructure not only in present but in future too (Zaldivar et al. 2020). The continuous analysis done by these security systems makes an accurate prediction about the present and possible future scenario in which the security of the IoT devices and infrastructure can be compromised by data hackers. It is, therefore, in this particular context of future attacks and understanding their modus operandi, IoT security analytics is important for the organization.

3.4.5 Testing Hardware

IoT hardware must undergo comprehensive and rigorous testing at regular intervals in order to make sure that no data is compromised. The testing of hardware is also required in order to make sure that the important parameters such as range, latency, and capacity of the devices are up to the mark. Take the example of a microprocessor chip which is an essential part of all IoT devices. It should be manufactured in a manner so that it offers optimum performance without compromising on safety and security aspect of the

IoT devices (Culbert 2020). It must be able to provide superior protection against data theft and hence it is the responsibility of the chip manufacturer to make sure all these desirable characteristics must be included while producing these microprocessors.

3.4.6 Developing Secure IoT Apps

One of the most effective measures that can help IoT devices and infrastructure to keep themselves safe from data hackers is the development of apps that comply with all safety features and security mechanisms. By developing such apps, which are safe and secure for use, data analytics firms will have an early advantage against data-hacking attempts that are made by hackers to get an entry into the security infrastructure of the company. It is also the responsibility of developers to integrate the security mechanisms in the app so as to make them immune from data theft and cyberattacks (Sohal et al. 2018).

3.4.7 Keeping Abreast with the Latest Security Threats

Information is power and keeping oneself abreast with the latest cybersecurity threats related to IoT is one of the credible ways to keep the ecosystem safe from the current and emerging threats of the future. It is also recommended that Data Analytics firms should avail the services of external consultants and cybersecurity experts so as to keep them safe from the emerging threat of data breaches and users' privacy (Thierer 2015). In addition, organizations must develop in-house capabilities that keep a close watch on the current scenario of cybersecurity and accordingly suggest measures and safety mechanisms for offering better safety solutions to IoT infrastructure.

3.5 Conceptual Framework for IoT Data Analytics

The conceptual framework proposed here (see Figure 3.5) consists of two different loops that must be connected through a continuous channel of feedback so as to make sure both IoT devices and infrastructure remains up to date and secure from the potential threats of cybercrimes and data hacking (Okikiolu 2021). Therefore, it is important to have a continuous loop of feedback that keeps on exchanging information to make sure both hardware and software components of IoT network are well equipped to avert the potential data breach (Stoter et al. 2021). The role of policymakers and regulators is also very important hers as their involvement in the process can really help to make the entire domain of IoT devices safe and secure from cybercrime and data breaches.

The rules and regulations related to cybersecurity are constantly evolving and even more stringent measures need to be introduced in order to make sure that hackers get deterred by the strong laws (Brumfield 2021). Data Analytics firms, on their part, need to become flexible and agile in their approach as the entire segment is changing at a constant pace. Hackers are constantly upgrading their methods to improve the data systems which means companies cannot afford to take a lenient and laid-back approach when it comes to tackling cybersecurity threats. There are implications for the individual users also, and they must use their discretion while sharing any kind of data or information with the third party.

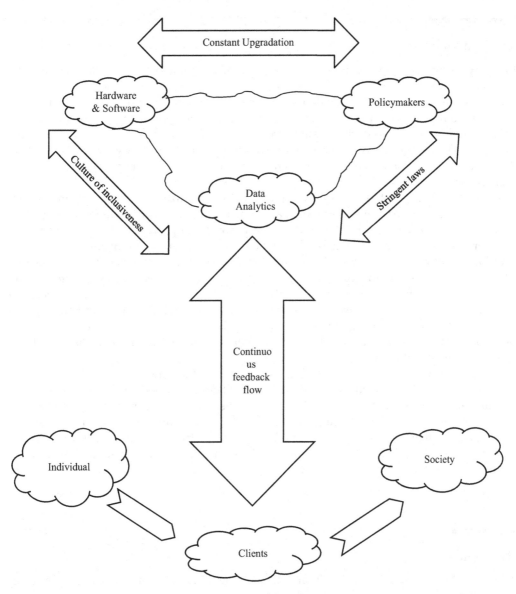

FIGURE 3.5
Conceptual framework for IoT security solutions for smart city applications: Author's own conceptualization.

Without exercising proper discretion and safety measures, there is a high probability that safety and security of users' data will get compromised and that could possibly result in both monetary and non-monetary implications for them. Hence, it is essential for the entire ecosystem of IoT devices to work in a mutually understandable manner and follow a holistic way of working that encourages each and every stakeholder to follow safety guidelines and protocols (Kore 2021).

3.6 Conclusion

In the wake of changing business realities, specifically related to the collection and dissemination of information, Data Analytics firms are increasingly grappled with enhanced safety and security challenges related to data of their clients and users. The threat of data hacking, compromised systems and infrastructure, infringement of integrity, and damaged reputation are some of the possible scenarios that each and every organization working in the Data Analytics domain wants to avoid. In fact, the business model of the Data Analytics firms is considerably dependent on the data safety and security they offer to their clients.

In such a scenario, it is absolutely important for organizations to make sure that the entire ecosystem of IoT devices works in a completely seamless and safe manner. In order to achieve this objective, firms need to adopt a holistic approach and should work in a manner that each and every stakeholder in the IoT domain understands the importance of providing safe and secure data collection and analysis services to the clients. Even one possible missing or weak link can jeopardize the entire safety mechanism adopted by the organization, and this is the reason why this chapter has suggested of holistic conceptual framework connecting all possible stakeholders involved in gathering and generation of data and information. The suggested framework also takes into consideration the important goal that needs to be played by policymakers and regulators to help the industry to grow at more enviable pace.

The statutory organizations throwing their weight behind the cybersecurity challenges in terms of more stringent rules and regulations will definitely bode well for enhancing the overall safety and security credentials of the ecosystem. The imposition of financial penalty along with criminal proceedings provisions against the culprits involved in cybersecurity crimes will go a long way in making sure that the domain of IoT devices and information-sharing infrastructure will remain secure from data-hacking threats in the future.

References

Alaba, F.A., Othman, M., Hashem, I.A.T., and Alotaibi, F. (2017), Internet of Things security: A survey. *Journal of Network Computers and Applications*, 88, 10–28.

Ali, S., Bosche, A., and Ford, F. (2018), *Cybersecurity Is the Key to Unlocking Demand in the Internet of Things*, Bain and Company, Boston, MA, USA.

Brumfield, C. (2021), Why today's cybersecurity threats are more dangerous. https://www.csoonline.com/article/3635097/why-today-s-cybersecurity-threats-are-more-dangerous.html (accessed on 5 October 2021).

Bugeja, J., Jacobsson, A., and Davidsson, P. (17–19 August 2016), On privacy and security challenges in smart connected homes. In *Proceedings of the European Intelligence and Security Informatics Conference (EISIC)*, Uppsala, Sweden, 172–175.

Culbert, D. (2020), Personal data breaches and securing IoT devices. https://betanews.com/2019/08/13/securing-iot-devices/ (accessed on 30 September 2021).

Ding, F., Song, A., Tong, E., and Li, J. (2016), A smart gateway architecture for improving efficiency of home network applications. *Journal of Sensors*, 10, 1–10.

Estrada, D., Tawalbeh, L., and Vinaja, R. (2020), How secure having IoT devices in our home. *Journal of Information Security*, 11, 81–91.

Gemalto (2020), Securing the IoT-building trust in IoT devices and data. https://www.gemalto.com/; https://www.gemalto.com/iot/iot-security (accessed on 29 September 2021).

Hassan, W.H. (2019), Current research on Internet of Things (IoT) security: A survey. *Computer Networks*, 148, 283–294.

Izzat, A., Chuck, E., and Loai, T. (2020), *The NICE Cyber Security Framework, Cyber Security Management*, Springer, Basel, Switzerland.

Khan, M.A. and Salah, K. (2018), IoT security: Review, blockchain solutions, and open challenges. *Future Generation Computer Systems*, 82, 395–411.

Khvoynitskaya, S. (2020), The history and future of the Internet of Things. https://www.itransition.com/; https://www.itransition.com/blog/iot-history (accessed on 02 October 2021).

Kore (2021), IoT security: The key to building trust. https://www.iotforall.com/iot-security-the-key-to-building-trust (accessed on 26 August 2021).

Liu, X., Zhao, M., Li, S., Zhang, F., and Trappe, W. (2017), A security framework for the internet of things in the future internet architecture. *Future Internet*, 9, 27.

Liyanage, M., Braeken, A., Kumar, P., and Ylianttila, M. (2020), *IoT Security: Advances in Authentication*, John Wiley & Sons, West Sussex, UK.

Meng, Y., Zhang, W., Zhu, H., and Shen, X.S. (2018), Securing consumer IoT in the smart home: Architecture, challenges, and countermeasures. *IEEE Wireless Communication*, 25, 53–59.

Monther, A.A. and Tawalbeh, L. (2020), Security techniques for intelligent spam sensing and anomaly detection in online social platforms. *International Journal of Electrical and Computer Engineering*, 10, 2088–8708.

Okikiolu, N. (2021), Using blockchain to improve security for IoT devices. https://internetofthings agenda.techtarget.com/post/Using-blockchain-to-improve-security-for-IoT-devices (accessed on 25 September 2021).

Singh, J., Thomas, F.J.-M., Pasquier, J.B., Ko, H., and Eyers, D.M. (2016), Twenty security considerations for cloudsupported Internet of Things. *IEEE Internet Things Journal*, 3, 269–284.

Sohal, A.S., Sandhu, R., Sood, S.K., and Chang, V. (2018), A cybersecurity framework to identify malicious edge device in fog computing and cloud-of-things environments. *Computer Security*, 74, 340–354.

Stoter, J., Ohori, K.A., and Noardo, F. (2021), Digital twins: A comprehensive solution or hopeful vision? https://www.gim-international.com/content/article/digital-twins-a-comprehensive-solution-or-hopeful-vision (accessed on 4 October 2021).

Tawalbeh, L.A. and Tawalbeh, H. (2017), Lightweight crypto and security. In: Houbing Song, Glenn A. Fink, Sabina Jeschke (Eds.), *Security and Privacy in Cyber-Physical Systems: Foundations, Principles, and Applications*, Wiley, West Sussex, UK, 243–261.

Tawalbeh, M., Quwaider, M., and Tawalbeh, L.A. (2020), Authorization model for IoT healthcare systems: Case study. In *Proceedings of the 2020, 11th International Conference on Information and Communication Systems (ICICS)*, Irbid, Jordan, 7–9 April, 337–342.

Thierer, A.D. (2015), The Internet of Things and wearable technology: Addressing privacy and security concerns without derailing innovation. http://jolt.richmond.edu/v21i2/article6.pdf (accessed on 01 October 2021).

Zaldivar, D., Tawalbeh, L., and Muheidat, F. (January 6 2020), Investigating the security threats on networked medical devices. In *Proceedings of the 2020, 10th Annual Computing and Communication Workshop and Conference (CCWC)*, Las Vegas, NV, USA, 0488–0493.

4

IOT-BDA Architecture for Smart Cities

Garima Pandey and C. Ramesh Kumar
Galgotias University, Greater Noida, India

Mayank Kumar
CEO of Readycoder Private Limited, Founder & CEO of Doubtfree EdTech Private Limited, Chief Operating Officer of Indo Biopearl Healthcare & Research, Chief Operating Officer of Foodenia, Noida, India

Kashish Gupta
Doubtfree EdTech Private Limited, Ghaziabad, India

Shivangi Singh Jha
Chief Business Development Officer, Doubtfree EdTech Private Limited, Chief Technology Officer, Foodenia, Noida, India

Jatin Jha
PR Manager of Doubtfree EdTech Private Limited, Noida, India

CONTENTS

4.1 Introduction 58
    4.1.1 What Is IoT? 59
    4.1.2 Prominent Advantages of IoT 60
    4.1.3 Smart Cities 60
4.2 Architecture of Smart City 61
    4.2.1 Level 1: Data Collection 61
    4.2.2 Level 2: Transit 61
    4.2.3 Level 3: Data Integration and Reasoning 62
    4.2.4 Level 4: Device Control and Alerts 62
    4.2.5 Smart Cities and Big Data Analytics 62
4.3 IoT in Smart Cities 63
    4.3.1 Use of IoT in Developing a Smart City 64
    4.3.2 Smart Lighting 64
    4.3.3 Smart Parking 64
    4.3.4 Smart Transit 65
    4.3.5 Smart Waste Management 66
    4.3.6 Smart Education 67
4.4 Applications of IoT in Smart Cities 68
    4.4.1 Why Do We Need BDA? 68
    4.4.2 Big Data Utilization in Information Management 69

DOI: 10.1201/9781003217404-4

4.4.3   Cloud Storage 70
4.4.4   Hybrid Data Storage 70
4.4.5   Security Framework in Smart Cities 71
4.5   Conclusion 72
4.6   Limitations and Future Research Extensions 72
References 72

4.1 Introduction

IoT is one of the most advanced and leading technology of this era, it is one of the most dynamic and advent technical advancement in the field of information and communication. Although communication technology has grown exponentially over the past two decades, it has recently been severely restricted from connecting common devices to end users, such as mainframes, desktop computers, laptops, smartphones as well as tablets [1]. In past years, we all have witnessed our increasing interaction and engagement with smart devices which made them a part of our life, and in this article, we are going to study about the same but on a whole another level because we will be discussing about smart cities or we can say cities with 100% implementation of IoT [2].

The first question may arise what does the word "Smart City" signify or what one should think of it when heard. Well, the answer to this is, not a universally accepted definition. It is perceived and interpreted differently by different people. As a result, smart city considerations differ from city to city and country to country, depending on the level of development, the readiness to change and the change itself, as well as citizen resources and aspirations. In India, the term "smart city" may have a different meaning than in Europe. There is no way to define the Wise City even in the middle of India [3]. All nations, including India, rely on cities to drive economic growth. According to Census 2011, 31% of India's population lives in cities, which account for 63% of the country's GDP. By 2030, urban regions are predicted to house 40% of India's population and contribute 75% of the country's GDP due to increased urbanization.

The concept of a smart city is modern in city development and upliftment. The smart cities aim at economic growth and improvement of people's quality of life by facilitating small-scale development and harnessing technology [4], Well, if we are talking about implementation of IoT in a whole city and as we know IoT means IoT and that also means all the data of the whole city has to be huge, messed up, or what we can say is "big". From here, arises the concept of BDA [5].

The BDA, or big data analytics, helps us to manage all the data, such as generous amount of random data that can be collected and stored in the cloud or in data centers using distributed error-tolerant information like Not SQL Only, to develop a single service or application. As a result, large-scale data processing planning models are used, and the same algorithms can be applied to data analysis gaining values on stored data. So yeah, we do need big data analysis, but after everything (BDA) big data analysis in smart media is still in an initial stage. But still big data analytics can be promising for the future of smart cities [6].

Big Data is a concept that works with data sets that are beyond the scope of manual conversion capabilities, and the IoT is gaining traction as business models recognize the need to use information technology to develop new processes and improve data management, knowing that Big Data is a concept that works with data sets that are beyond the scope of manual manipulation techniques. This chapter was prepared because there is a lack of research on smart cities and how they are expected to and will work utilizing IoT in conjunction with big data analytics [7].

This article is useful for anyone who is interested in learning more about smart cities, as it focuses on the analysis of large data applications, which is seen as a major issue in the smart cities sector. Finally, accessible challenges will be introduced in order to provide additional clarity on the direction of future research in the smart city big data field. This chapter will look at how big data is used in smart cities [8, 9].

4.1.1 What Is IoT?

According to the ITU (International Telecommunication Union), "Internet of Items" is a broad word that can be used to describe anything connected to the Internet. Over time, however, the term "Internet of Things" has become associated with "conversational" things [10]. In terms of smart cities, The IoT is built on three fundamental functions of smart objects: traceability, communication, and interaction, as depicted in Figure 4.1.

It points us to a vast network of self-contained digital machines that have an impact on our daily lives [11]. Among the equipment that can detect, analyze, and control numerous aspects of city life are intelligent sensors, monitoring devices, AI systems, and actuators. Many sensors, for example, can collect meteorological data that can be used to manage thermostats in public buildings, lowering carbon emissions and saving money for the city [12].

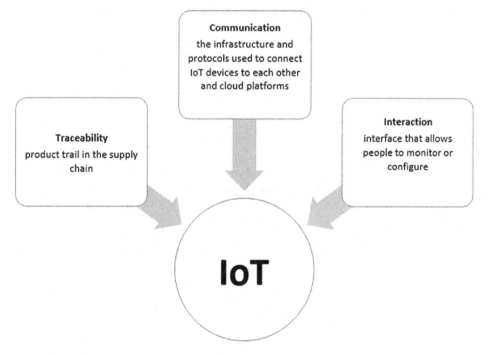

FIGURE 4.1
IoT for smart city.

4.1.2 Prominent Advantages of IoT

As now you are well aware about the concept of IoT, let's now understand what are the significant advantages of IoT in detail.

- Enhance customer experience
- Enhance security and safety in business
- Reduces/decreases operational cost
- Gather data for business decisions

As now you have come to know about what IoT means and its advantages, let's now discuss about how to make a city an IoT controlled smart city as the prime concern of this book is basically to draw the attention of the reader about the same [13].

4.1.3 Smart Cities

A wise city is a collection of concepts that cover economic, social, government, travel, environmental, and lifestyle., Public transportation, traffic analysis, resource monitoring, environmental monitoring, and incident reporting are all part of this theory. The data collected in the above- mentioned areas assists city officials in improving infrastructure and improving their facilities as shown in Figure 4.2.

Its mission is to improve the efficiency and effectiveness of resource management while also providing services to meet the general and specific needs of the community [12].

According to polls and publications, the government does not mandate any particular model for smart cities to follow. There is no such thing as a "one-size-fits-all" solution; thus

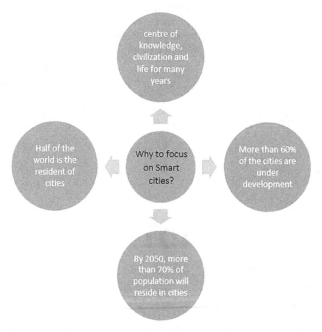

FIGURE 4.2
Infrastructure for smart city.

each city must choose and develop its own smart city idea, vision, goal, and plan (proposal) that is relevant to its local context, resources, and levels of ambition. As a result, everyone should answer the question to determine their smart city model. What kind of smart city do citizens desire? To do so, communities must create their own Smart City Proposal (SCP), which includes a vision, resource integration plan, targeted infrastructure, and smart application outcomes [14]. The smart city, also known as an eco-city or a sustainable city, strives to increase urban service quality while lowering expenses. It stands out due to its distinguishing characteristics, which include savvy management, lifestyle, mobility, housing, and a smart economy [15].

Their primary purpose is to match technological advancements to the city's future economic, social, and environmental concerns. Their appeal for togetherness is for a better quality of life. Most people prefer to live in cities as a result of urbanization and industrialization since they are the center of resources, information, and innovation. According to projections, 70% of the world's population will be urbanized by 2050, resulting in increasing demand for infrastructure and modern technologies to maintain a respectable standard of living. Smart city architecture is the solution to this since it gives "smart solutions" such as smart health, smart environment, smart energy, smart security, smart office, smart administration, smart transportation, and smart industries [16, 17].

4.2 Architecture of Smart City

We are excited about the future of smart city programs that will provide cities with stronger, smarter, and more flexible support, thanks to wireless technology and wireless sensor networks [18].

The sensor in the first level offers raw data for the generation of information. The sensor's data is acquired with the use of existing communication resources. PS devices, for example, acquire raw data through a satellite network and the Internet. Semantic web technologies and Dempster–Shafer combination rules are used to handle and evaluate the obtained data. This design is useful for completing daily chores or intelligently supporting people, as alerts and warnings can be used to remind end users of their tasks or monitor people's health while performing an activity [15, 19].

Life will become so easy after the utilization of smart city architecture. Let's now understand each level in detail as shown in Figure 4.3.

4.2.1 Level 1: Data Collection

The sensors present in the initial layer provide the primary data through which Information is generated. Discrete data is of csv or text message format. Processing of the collected formats into common format is done using semantic web technologies. This conversion is done in the next level [20].

4.2.2 Level 2: Transit

The data gathered at the previous level is evaluated at higher levels in order to make it web-accessible using semantic web technologies. The primary goal of Transit is to organize the data collected, for example, using the Resource Description Framework (RDF), which is utilized in a variety of intelligent software activities [21].

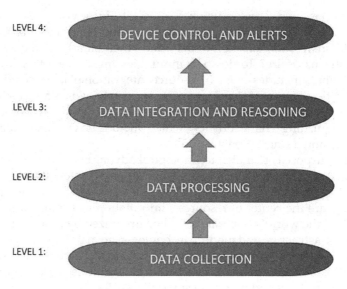

FIGURE 4.3
The utilization of smart city architecture.

4.2.3 Level 3: Data Integration and Reasoning

Data is converted into common format and then enhanced with OWL concepts with the help of the knowledge of domain experts.

Web ontology language (OWL) is an RDF graph representation used for publishing ontologies. Relationships are formed using the data and object properties. Knowledge can be further augmented by domain experts and uncertain reasoning once classification is complete. Dempster–Shafer is used to recognize activity and learn new rules in a certain area. The new principles discovered when extracting high-level context information from raw sensor data can subsequently be used to instill knowledge in smart city architecture [22].

4.2.4 Level 4: Device Control and Alerts

Multiple web applications can use prior level information to create intelligent working circumstances. The resulting data can be used for a variety of purposes, including input/output, messaging, alerts, and warnings.

4.2.5 Smart Cities and Big Data Analytics

The population is increasing day by day and because of that they require services that meet their daily requirements. Over the period of time, we have seen increase in the demand for embedded devices such as smartphones, sensors, and digital cameras as these devices become a gateway to communicate with each other over the Internet. An IoT-based system has great potential in the development of smart cities and urban planning.

Smart Metropolis is a well-developed city that includes smart domains. Rudolf Giffinger's seminal work outlined six smart domains that characterize the smartness of a smart city [23]. As shown in Figure 4.4, smart cities contain clever people, a high level of living, the

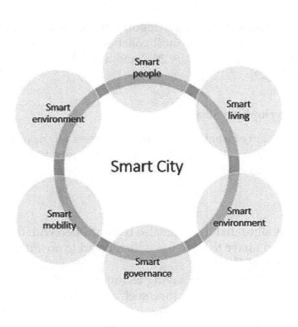

FIGURE 4.4
Smart city domains.

economy, government, transportation, and a smart ecology. While scholars disagree on the number and nature of the intelligent settings that make up the SC, there is widespread agreement that city intelligence is reflected in domain relations and information sharing [24]. It is intuitively possible to achieve this by leveraging ICT-based solutions.

ICT generates a plethora of large data volumes in a variety of industries. Big Data is the term for these large amounts of data. It consists of structured, semi-structured, and unstructured information [25, 26]. Similarly, Big Data Analysis is the name given to the process of analyzing large amounts of data. Extraction of insights and hidden connections has gotten a lot of attention in the IS to help with decision-making [1]. The applicability of smart cities and Big Data Analysis has sparked debate in the literature, with various viewpoints. We can identify the amounts of obtained Big Data throughout various stages due to the complex and dynamic character of SC projects and the uniqueness of the engaged stakeholders from the beginning stages of planning to the final stages of periodic service management and monitoring. Despite the fact that there are numerous papers addressing the adoption of BDA in various SC sectors, only a few of them have proposed developing sector-specific BDA frameworks that may serve the analytical needs at the countable project stage of SC [27].

4.3 IoT in Smart Cities

The IoT technology has been a crucial pillar for the development of smart cities since the notion was originally established. As technology advances and more countries embrace next-generation communication, IoT technology will continue to flourish and have a significant impact on how we live [28]. The Software-Defined Network (SDN) anticipates that there will be more than 75.44 billion linked IoT devices by 2025, according to advanced

Internet security (IoT) surveys. According to reports and current reports and trend research, the IoT is likely to grow into one of the most collaborative program in history, with more than 7.33 billion mobile users predicted by 2023 and more than 1,105 million mobile device users predicted by 2022 [29].

4.3.1 Use of IoT in Developing a Smart City

In ancient times, the concept of smart city would have been a topic of joke, but now it is the topic of everyone's attention. Let's learn about some cases where the concept of IoT can be used for developing a smart city.

4.3.2 Smart Lighting

Nobody would have thought that light can also be controlled with the help of smart technology where people don't have to go to the switch board to on–off the light. But now it is possible with the help of IoT [30, 31].

It is a remotely controlled resource saving technique by the means of which one can control the light on the basis of some factors, such as weather condition, time of the year, and detection of the motion as depicted in Figures 4.5 and 4.6. According to a research, 20% of energy of the public light is accumulated for use by Barcelona City [32].

4.3.3 Smart Parking

The technology and the standard of living have developed so much that the parking of vehicles has also [33] become easier by the means of techniques developed for smart parking with the help of concepts of IOT as depicted in Figure 4.7.

Resources can be saved by the means of Smart Parking technology. The smart parking technique saves our time by providing us with the parking spot. As the parking time is reduced fuel/gas is burnt less in searching of parking spot as a result of which fuel is burnt less and CO_2 is also emitted in less amount. Smart parking also reduces traffic rules violation as it detects the parking slots available and also routing the drivers to the spot by the means of an app [34, 35].

In the city of San Francisco, parking search time was reduced by 43% which not only saved time but also helped in saving resources (fuel usage). CO_2 emissions were also reduced by 30%. A research shows that there was a decrease in the number of parking violations and citations by 23%. RFID and [36, 37] IOT-based car parking system can be used

FIGURE 4.5
Smart lighting.

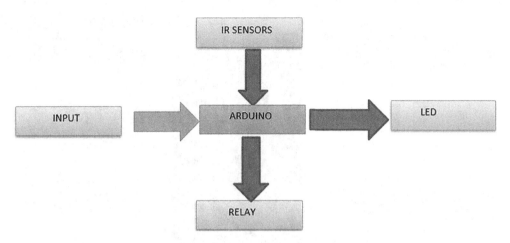

FIGURE 4.6
Smart lighting and application.

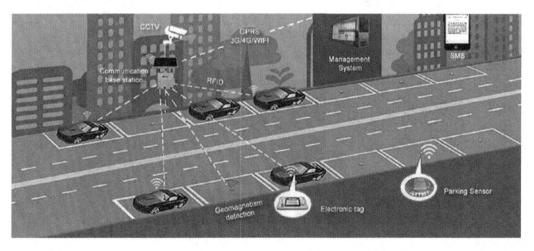

FIGURE 4.7
Smart parking technology.

at all parking areas mainly companies, industrial areas, and commercial area as depicted in Figure 4.8 [38].

4.3.4 Smart Transit

Customer service and safety can be increased by the means of smart transit. It provides a secure and safe ride by the means of providing ridership with video surveillance [39]. Real-time vehicle location tracking, the state of its static and dynamic motion can also be provided by the help of smart transit as shown in Figure 4.9. It increases the level of customer service by providing services like Wi-Fi, entertainment, and real-time program (set up) [40].

In Bucharest, the level of ride has increased because people are using this technology to have a secure, safe, and enjoyable journey. Smart Transit is also beneficial for visually impaired person because it provides a signal in the app whether the vehicle is approaching

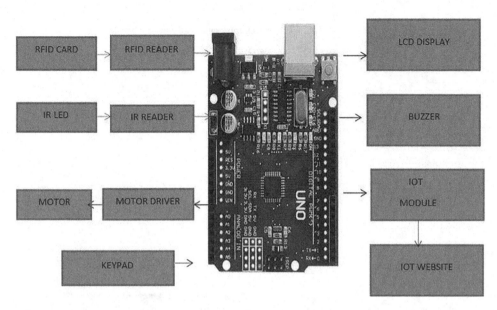

FIGURE 4.8
Block diagram of IoT-based parking system

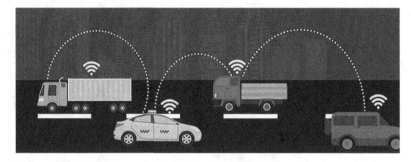

FIGURE 4.9
Real-time vehicle location tracking.

or not. It is also beneficial for drivers because it provides them the information whether a passenger needs their abetment in boarding at a stop or location.

4.3.5 Smart Waste Management

The management and disposal of waste can also administer by the help of IOT-based smart waste system as it uses sensors to notify the level of waste present in the dustbins and then communicate it to the management authority as shown in Figure 4.10 [41].

The following elements can help in the reduction of smart bin as its deployment in Australia decreases 75% of waste. It provides real-time access to provide dustbin fill-level information and enhances waste collection with smarter routes [42]:

- Fleets
- Fuel usage
- CO_2 emission

4.3.6 Smart Education

Nobody would have imagined studying without the need of going to schools physically, but the advancement in technology has played a major role in making it possible and feasible method of teaching [43, 44]. Last year the world witnessed a very deadly pandemic-COVID 19 due to which everything was shut down all around the world. Teaching was made possible by the means of smart education in Figure 4.11.

FIGURE 4.10
Smart waste management for IOT.

FIGURE 4.11
Smart education.

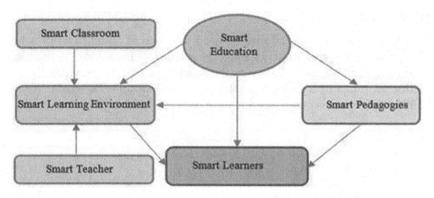

FIGURE 4.12
Smart teaching learning.

There are endless opportunities and challenges that one can explore in the field of smart education. You must be wondering how one can determine whether education is smart education or not, or in another way how to define smart education [45].

Different people have provided different definitions for the same. According to Zhu and He "smarter education aims to create intelligent environments by using smart technologies" [46], so that clever teaching can be simplified as providing personal learning services and empowering students to develop intellectual talents with a better value, higher thinking quality, and stronger moral ability as described in Figure 4.12. Another definition of smart education was provided by BAJAJ and SHARMA, which is "about providing personal education, anywhere and anytime".

4.4 Applications of IoT in Smart Cities

- Smart transport, energy, and health resources and hygiene
- Assistance in crime prevention
- Social governance and security
- Monitoring and maintenance of infrastructure including disaster management and emergencies
- Smart citizens
- Tourism
- Recreation and environmental management

To implement IoT in smart cities, we need Big Data Analysis (BDA). You must be intrigued in knowing about BDA.

4.4.1 Why Do We Need BDA?

Intelligent solutions can be created using connected technology and big data that can help to solve problems, improve life quality of urban dwellers, and reduce utilization of resources. The IoT is a necessary component for full functioning of a true smart city.

However, when working with such systems, a new set of models and analytical methods will be required, as they are largely based on traditional analysis, the only data-driven methods used in traditional Big Data systems (e.g., commercial or public data). We invite experts from various fields to discuss the latest Big Data research, IoT, and specific industry.

Big Data is a term that refers to the accumulation of complex and sophisticated data from sensors, social media, applications, and devices that are connected to the Internet and require the use of advanced technology to store, manage, analyze, and visualize data. Because it creates a global revolution in the use of media and technologies, Big Data Analysis (BDA) is a marketing strategy. The BDA is utilized to give better data management in real time, as data management will generate support systems that allow for intelligent system integration. Big Data now allows scientists and advertisers to study positive aspects of human existence by taking into account the amount of data created on a daily basis and assessing its complexity. Its role is to turn the data into tools that management can use to solve problems and make well-informed decisions.

4.4.2 Big Data Utilization in Information Management

Big data usage encompasses all business objectives that require data access, analysis, and integration in business decision-making. This chapter provides an overview of big data applications, with a focus on decision-making using big data in a variety of industries. Because it is connected to cloud environments with proper data flow, Big Data gains power to information management. This responds to smart city challenges, which include big data and the need for efficient systems (speed, variability, authenticity, data transfer within Big Data). In recent years, accurate data analysis has led to a concerted effort between academics and industry to find ways to reduce energy consumption, traffic congestion, and air pollution in Figure 4.13.

Big data has the following impacts on different perspectives that are mentioned below:

- Public security:

  Big data plays an important role in public safety. Any smart cities big data analysis can be used to detect the crimes or we can say to protect the crimes in the cities. Both IoT and big data analysis help in detecting exact crime location.

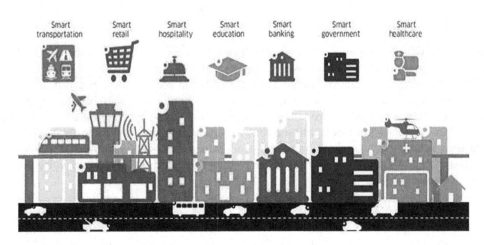

FIGURE 4.13
Big data utilization in IOT.

- Transportation:
  As we all know, time is important to everyone either he/she is a businessman or a student, big data analysis plays a major role by analyzing data from transport authorities. Sensors in the vehicles analyze the data and help in reducing traffic congestion.

- Reduction of cost:
  Big data analysis suggests which areas are required in a smart city for the transformation, or we can see transforming a city into a smart city, it also provides what kind of transformation is needed. This can make cities to invest for required areas.

- Various storage in smart cities:
  Before coming to the storage of the data in smart cities, firstly, we have to know about some data used for data analysis. Smart cities require a large amount of real time data to function efficiently.

- Smart cars data:
  Big data analysis and IoT plays important role in making devices smart and autonomous, especially smart cars, which are becoming more integrated with mobile systems. This helps us to detect the location of the car and prevent road accidents, provide shortest path to reach our distant destination. A well-known example of this is Tesla car which has an auto drive feature and many other things which makes it a smart car.

- Camera System Data:
  Camera surveillance is the most usable device in our daily lives. It is mostly used for security purposes, and nowadays these cameras are more advanced using IoT, now it can serve as a crime detector as well as prove public safety, build intelligent traffic systems.

- Environmental sensor Data:

  There are many sensors of IoT like air quality sensors which help cities to detect and take action against polluters and also help us to locate the non-green areas where planting of trees is important. This helps us to suggest plant the trees for better oxygen level. All smart cities store their data mainly in these locations:

4.4.3 Cloud Storage

As we all know, solid state drives are faster than hard drives, and we need to store large amount of data for analytical purposes. Cloud systems use these drives in their data centers; these systems have an advantage of removing data redundancy, and it helps in the encryption of transmitted data. Usually, cloud services or cloud storage is better option for IOT devices. Using cloud storage is the safest and the easiest way to use and store your data and analyze it to get information about a particular thing. Just like nowadays there are various smart watches which have this heart rate sensor, sleep monitoring which synchronizes to our smart phones and saves our data to the clouds so that whenever we want, we can access it.

4.4.4 Hybrid Data Storage

Hybrid data storage systems are the combination of benefits of both clouds as well as edge storage. In the edge storage, it helps to process the data that is close to the source and is cheaper than streaming data to remote storage location, basically it uses artificial intelligence which is already under development in cities; it automatically detects traffic and accidents and according to that it provides faster way of response to different situations.

The main rule of big data is two processes that data from IoT devices as well as sensors recognize the patterns as well as needs; and this analysis helps us to reduce the number of problems which we are facing in daily lives, for example road accidents and parking spot detection; it can also reduce crime by improving smart urban lightning and many energy-related systems.

4.4.5 Security Framework in Smart Cities

Having security in smart cities is the topmost priority and everything is IoT based so everywhere is having a technology and securities are necessary for the efficient way of working of any IOT devices and to prevent the leakage of data. Just like the other technologies which can be hacked, IOT devices are also vulnerable to various security threats; and it cannot be compromised in the smart cities as shown in Figure 4.14. As we all know, firewalls play some most important role in securing devices and different systems may have physical or virtual firewalls of their own, and nowadays blockchain technology is also used with IoT so that it increases the security of IOT devices. Figure 4.14 shows security architecture in smart cities.

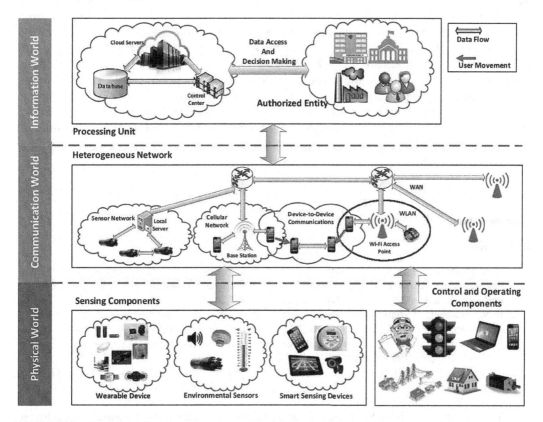

FIGURE 4.14
Architecture for smart city: physical world, communication world, and information world.

4.5 Conclusion

Since the notion of smart cities was first introduced, the IoT has been a fundamental component of smart city development. IoT will undoubtedly have a significant impact on our way of life in the future. Many countries are now attempting to integrate next-generation technologies for the benefit of their citizens. Along with Big Data Analytics, the IoT is predicted to grow into one of the smartest collective and collaborative systems in history, providing tremendous possibilities across a wide range of sectors, including urban mobility, healthcare, security, and management. The ability to quickly evaluate large data sets and identify opportunities for efficiency, productivity, and profitability. Interconnectivity is one of the most important building blocks of next-generation smart city development. Citizens and governments will be linked in never-before-seen ways. IoT will provide smart cities with many chances and benefits, but it will also present certain obstacles.

4.6 Limitations and Future Research Extensions

Even though IoT in smart cities and Big Data Analysis have great advantages, but along with that there are limitations which have to be worked upon in the future for smooth functioning and usage of next-generation technology. These limitations include joining various general-purpose data analysis engines like Anaconda, TensorFlow, etc., Security and Privacy, Lack of encryption, etc. Currently, we are implementing IoT Technology, but it lacks perfection as there are ways that the interlocking systems of our homes which we used for our protection and comfort can be hacked or malfunctioning of the smart parking system used to spot a parking location will lead to time wastage and delay in repairing the system as IoT technology is not well versed or designed to analyze the defect and make amendments, so we can't fully rely on IoT. A lot of research and trial will help to eliminate any shortcoming of this technology so that it can be used in the future in the best possible way to help humans.

References

1. S. A. Shah, D. Z. Seker, M. M. Rathore, S. Hameed, S. Ben Yahia, and D. Draheim, "Towards disaster resilient smart cities: Can Internet of Things and big data analytics be the game changers?," *IEEE Access*, 2019, doi: 10.1109/ACCESS.2019.2928233.
2. B. N. Silva et al., "Urban planning and smart city decision management empowered by real-time data processing using big data analytics," *Sensors (Switzerland)*, 2018, doi: 10.3390/s18092994.
3. J. Han, Y. Yoon, and K. Han, "Urban planning and smart city decision data processing using big data analytics," *Sensors*, 8, no. 9 (2018): 2994.
4. S. Chakrabarty and D. W. Engels, "A secure IoT architecture for smart cities," In *2016 13th IEEE annual consumer communications & networking conference (CCNC)*, pp. 812–813. IEEE, 2016, doi: 10.1109/CCNC.2016.7444889.
5. Y. Simmhan, P. Ravindra, S. Chaturvedi, M. Hegde, and R. Ballamajalu, "Towards a data-driven IoT software architecture for smart city utilities," *Softw. Pract. Exp.*, 2018, doi: 10.1002/spe.2580.

6. H. A. Kumar, J. Rakshith, R. Shetty, S. Roy, and D. Sitaram, "Comparison of IoT architectures using a smart city benchmark," *Procedia Com. Sci.* 171 (2020): 1507–1516, doi: 10.1016/j.procs.2020.04.161.

7. K. Szum, "IoT-based smart cities: A bibliometric analysis and literature review," *Eng. Manag. Prod. Serv.*, 2021, doi: 10.2478/emj-2021-0017.

8. N. Gavrilović and A. Mishra, "Software architecture of the internet of things (IoT) for smart city, healthcare and agriculture: analysis and improvement directions," *J. Ambient Intell. Humaniz. Comput.*, 2021, doi: 10.1007/s12652-020-02197-3.

9. P. Marques et al., "An IoT-based smart cities infrastructure architecture applied to a waste management scenario," *Ad Hoc Networks*, 2019, doi: 10.1016/j.adhoc.2018.12.009.

10. M. Geetha Pratyusha, Y. Misra, and M. Anil Kumar, "IoT based reconfigurable smart city architecture," *Int. J. Eng. Technol.*, 2018, doi: 10.14419/ijet.v7i2.7.10287.

11. P. Ta-Shma, A. Akbar, G. Gerson-Golan, G. Hadash, F. Carrez, and K. Moessner, "An ingestion and analytics architecture for IoT applied to smart city use cases," *IEEE Internet Things J.*, 2018, doi: 10.1109/JIOT.2017.2722378.

12. I. Miladinovic, S. Schefer-Wenzl, and H. Hirner, "IoT architecture for smart cities leveraging machine learning and SDN," In *2019 27th Telecommunications Forum (TELFOR)*, pp. 1–4. IEEE, 2019, doi: 10.1109/TELFOR48224.2019.8971033.

13. B. N. Silva, M. Khan, and K. Han, "Towards sustainable smart cities: A review of trends, architectures, components, and open challenges in smart cities," *Sustain. Cities Soc.*, 2018, doi: 10.1016/j.scs.2018.01.053.

14. I. Marcu, G. Suciu, C. Bălăceanu, A. Vulpe, and A. M. Drăgulinescu, "Arrowhead technology for digitalization and automation solution: Smart cities and smart agriculture," *Sensors (Switzerland)*, 2020, doi: 10.3390/s20051464.

15. M. Cerchecci, F. Luti, A. Mecocci, S. Parrino, G. Peruzzi, and A. Pozzebon, "A low power IoT sensor node architecture for waste management within smart cities context," *Sensors (Switzerland)*, 2018, doi: 10.3390/s18041282.

16. M. Fahmideh and D. Zowghi, "IoT smart city architectures: An analytical evaluation," *Information Systems* 87 (2020): 101409. doi: 10.1109/IEMCON.2018.8614824.

17. T. Shaikh, S. Ismail, and J. D. Stevens, "Aura Minora: A user centric IOT architecture for smart city," In *Proceedings of the International Conference on Big Data and Advanced Wireless Technologies*, pp. 1–5. 2016, doi: 10.1145/3010089.3016028.

18. S. El Khateeb, "IoT architecture a gateway for smart cities in Arab world," In *2018 15th Learning and Technology Conference (L&T)*, pp. 153–160. IEEE, 2018, doi: 10.1109/LT.2018.8368500.

19. A. Kumari, R. Gupta, and S. Tanwar, "Amalgamation of blockchain and IoT for smart cities underlying 6G communication: A comprehensive review," *Comput. Commun.*, 2021, doi: 10.1016/j.comcom.2021.03.005.

20. D. Li, Z. Cai, L. Deng, and X. Yao, "IoT complex communication architecture for smart cities based on soft computing models," *Soft Comput.*, 2019, doi: 10.1007/s00500-019-03827-5.

21. J. ho Park, M. M. Salim, J. H. Jo, J. C. S. Sicato, S. Rathore, and J. H. Park, "CIoT-Net: a scalable cognitive IoT based smart city network architecture," *Human-centric Comput. Inf. Sci.*, 2019, doi: 10.1186/s13673-019-0190-9.

22. G. Kaur, P. Tomar, and P. Singh, "Design of cloud-based green IoT architecture for smart cities," In *Internet of Things and Big Data Analytics Toward Next-Generation Intelligence*, pp. 315–333. Springer, Cham, 2018.

23. K. Soomro, M. N. M. Bhutta, Z. Khan, and M. A. Tahir, "Smart city big data analytics: An advanced review," *Wiley Interdiscip. Rev.: Data Min. Knowl. Discov.*, 2019, doi: 10.1002/widm.1319.

24. G. Sion, "Smart city big data analytics: Urban technological innovations and the cognitive internet of things," *Geopolit. Hist. Int. Relations*, 2019, doi: 10.22381/GHIR112201910.

25. A. M. S. Osman, "A novel big data analytics framework for smart cities," *Futur. Gener. Comput. Syst.*, 2019, doi: 10.1016/j.future.2018.06.046.

26. K. Susmitha and S. Jayaprada, "Smart cities using big data analytics," *Int. Res. J. Eng. Technol.*, 2017.

27. B. N. Silva, M. Khan, and K. Han, "Integration of Big Data analytics embedded smart city architecture with RESTful web of things for efficient service provision and energy management," *Futur. Gener. Comput. Syst.*, 2020, doi: 10.1016/j.future.2017.06.024.

28. Z. Khan, A. Anjum, K. Soomro, and M. A. Tahir, "Towards cloud based big data analytics for smart future cities," *J. Cloud Comput.*, 2015, doi: 10.1186/s13677-015-0026-8.

29. M. Fugini, J. Finocchi, and P. Locatelli, "A big data analytics architecture for smart cities and smart companies," *Big Data Res.*, 2021, doi: 10.1016/j.bdr.2021.100192.

30. I. Chew, D. Karunatilaka, C. P. Tan, and V. Kalavally, "Smart lighting: The way forward? Reviewing the past to shape the future," *Energy Build.*, 2017, doi: 10.1016/j.enbuild.2017.04.083.

31. G. Pasolini, P. Toppan, F. Zabini, C. D. De Castro, and O. Andrisano, "Design, deployment and evolution of heterogeneous smart public lighting systems," *Appl. Sci.*, 2019, doi: 10.3390/app9163281.

32. B. Sun, Q. Zhang, and S. Cao, "Development and implementation of a self-optimizable smart lighting system based on learning context in classroom," *Int. J. Environ. Res. Public Health*, 2020, doi: 10.3390/ijerph17041217.

33. J. J. Barriga et al., "Smart parking: A literature review from the technological perspective," *Appl. Sci. (Switzerland)*, 2019, doi: 10.3390/app9214569.

34. T. Perković, P. Šolić, H. Zargariasl, D. Čoko, and J. J. P. C. Rodrigues, "Smart parking sensors: State of the art and performance evaluation," *J. Clean. Prod.*, 2020, doi: 10.1016/j.jclepro.2020.121181.

35. S. Alkhuraiji, "Design and implementation of an android smart parking mobile application," *TEM J.*, 2020, doi: 10.18421/TEM94-06.

36. F. Al-Turjman and A. Malekloo, "Smart parking in IoT-enabled cities: A survey," *Sustain. Cities Soc.*, 2019, doi: 10.1016/j.scs.2019.101608.

37. W. Z. Al Qaidhi and M. Sohail, "Smart parking system using IOT," *J. Student Res.*, 2020, doi: 10.47611/jsr.vi.881.

38. Y. Agarwal, P. Ratnani, U. Shah, and P. Jain, "IoT based smart parking system," In *2021 5th International Conference on Intelligent Computing and Control Systems (ICICCS)*, pp. 464–470. IEEE, 2021, doi: 10.1109/ICICCS51141.2021.9432196.

39. H. Cao and M. Wachowicz, "The design of an IoT-GIS platform for performing automated analytical tasks," *Comput. Environ. Urban Syst.*, 2019, doi: 10.1016/j.compenvurbsys.2018.11.004.

40. A. Bashir and A. H. Mir, "Secure framework for internet of things based e-health system," *Int. J. E-Health Med. Commun.*, 2019, doi: 10.4018/IJEHMC.2019100102.

41. P. S. A. Mahajan, A. Kokane, A. Shewale, M. Shinde, and S. Ingale, "Smart waste management system using IoT," *Int. J. Adv. Eng. Res. Sci.*, 2017, doi: 10.22161/ijaers.4.4.12.

42. A. A. Selvaraj, "Smart waste management using IoT," *Int. J. Res. Appl. Sci. Eng. Technol.*, 2019, doi: 10.22214/ijraset.2019.3015.

43. S. Chaiyarak, A. Koednet, and P. Nilsook, "Blockchain, IoT and fog computing for smart education management," *Int. J. Educ. Inf. Technol.*, 2020, doi: 10.46300/9109.2020.14.7.

44. A. H. K. Mohammed, H. H. Jebamikyous, D. Nawara, and R. Kashef, "IoT text analytics in smart education and beyond," *J. Comput. High. Educ.*, 2021, doi: 10.1007/s12528-021-09295-x.

45. S. Chawla, D. P. Tomar, and D. S. Gambhir, "Smart education: A proposed IoT based interoperable architecture to make real time decisions in higher education," *Rev. Gestão Inovação e Tecnol.*, 2021, doi: 10.47059/revistageintec.v11i4.2589.

46. S. Jain and D. Chawla, "A smart education model for future learning and teaching using IoT," In *International Conference on Information and Communication Technology for Intelligent Systems*, pp. 67–75. Springer, Singapore, 2020, doi: 10.1007/978-981-15-7062-9_7.

5

Intelligent Framework for Smart Traffic Management System: Case Study

K. Aditya Shastry
Nitte Meenakshi Institute of Technology, Bengaluru, India

H. A. Sanjay
M.S. Ramaiah Institute of Technology, Bengaluru, India

M. Lakshmi
Nitte Meenakshi Institute of Technology, Bengaluru, India

CONTENTS

5.1   Introduction 75
    5.1.1   Internet of Things (IoT) in Intelligent Transport Systems 76
    5.1.2   Big Data Analytics and Its Role in Traffic Management Systems 78
5.2   Technologies Used in Intelligent Traffic Management Systems 78
5.3   Basic Working of an ITMS 81
5.4   Case Studies on ITMS 81
    5.4.1   ITMS in India 81
    5.4.2   Exigency Ambulance Management System 82
    5.4.3   Case Study-3 84
    5.4.4   Case Study-4 87
        5.4.4.1   Geographic Chart Information Handling 88
        5.4.4.2   Automobile Recognition and Physical Size Assessment 89
        5.4.4.3   Lane Occupancy and Increasing Queues 89
        5.4.4.4   "Display" Alert Communications 89
5.5   Research and Applications of ITMS 90
5.6   Challenges in Smart City Applications 94
5.7   Summary 95
5.8   Conclusion and Future Scope 96
References 96

5.1 Introduction

The development of faster computation devices has enabled us to use innovative solutions to every problem with the help of software. The ITMS is one such application software that enables us to solve traffic-related issues by providing services to different modes

DOI: 10.1201/9781003217404-5

of transport. It provides the users with information that users can utilize to make better decisions and hence be safer, more efficient, and smarter. In ITMS, transmission and data technologies are utilized in the domain of highway transport, road transportation, automobiles, consumers, and transport administration. The end users can interact with the functional interface of ITMS. The interface also provides information on additional forms of transportation to enhance the effectiveness of vehicles on roads. Attention in ITMS is increasing quickly because of growing matters linked to interior safety, as "ITMS" comprises inspection of highways, which is a crucial necessity in the domain of internal protection. In cases of unfortunate events such as fires, terrorist attacks natural disasters, etc., ITMS comes in very handy to evacuate people from such situations.

In developing nations such as India, movement from the countryside to urban environments because of quick urbanization and development is causing densely populated areas devoid of substantial infrastructural growth of the neighborhoods. Big Cities, like "New Delhi" and "Mumbai", are the most impacted cities. With the increase in population increase in usage of transport is significant. People use different modes of transportation, from bicycles, auto-rickshaws to metros and trains. This is leading to an exponential increase in road traffic. This increase, in turn, creates transport congestion and clogs, leading to a rise in the price of transport and influencing the everyday lives of individuals. Along with these difficulties like a waste of energy and time, rising ecological contamination, disasters, and instances of road anger are also emerging. Numerous additional causes are triggering this abrupt surge in highway transportation, including an increase in the populace (causing an upsurge in the number of automobiles on the highway), inadequate highway size, transportation regulator "lights", non-accessibility of comprehensive and appropriate data concerning transportation density on diverse paths, ineffective transportation organization and uncontrolled request for automobiles.

According to this scenario, the characteristic of transportation makes it hard to assess the highway transportation density on a live basis, to create improved transport-linked choices, and control the transportation effectively. Due to this, cities are not capable of resolving these issues. While the focus is on constructing highways, crossings, and passageways, and establishing alternate forms of bulk municipal transportation structures, this is inadequate to tackle traffic crowding and administration, which is becoming challenging. Consequently, there is a pressing necessity for establishing a current method for solving these problems. The resolution lies in leveraging innovative tools and smart results by implementing ITMS. This approach can trace the movement and speed of automobiles to deliver live transport administration, which is more active and cooperative of the fluctuating nature of transportation intensity [1].

5.1.1 Internet of Things (IoT) in Intelligent Transport Systems

The concept of a "smart city" relies on the tools adopted to enhance the citizen's value of life. The intelligent urban authority is a few major facets of "smart city" programs that will enable the organizing methods for improved judgment making [2–5]. One of the crucial aspects of the "smart city" control structure is the community value created out of the intelligent facilities offered [6].

Smart city solutions must include solutions for intelligent medical treatment, supervision, transportation supervision, parking solutions, transport, etc. to produce community importance for the assistance they offered. The beginning of the "internet of things (IoT)" has developed the notion of "smart cities". In a "smart city" ecosystem, the natural organizations of the municipality are armed with intelligent machines that constantly deliver multidimensional information in diverse areas, and this information is managed to accomplish smartness for the organization [6]. Eventually, smartness is employed to enhance the socio-financial actions of society.

Intelligent transportation structure is a vital element of "smart city" programs since transportation crowding is a serious concern that intensifies with urban growth. Intelligent transportation administration incorporates smart transportation methods with cohesive elements such as customized transportation signal regulations, expressway administration, crisis supervision facilities, and pavement modules [7]. Such methods gather live transportation information and take essential actions to prevent or reduce any public concern generated due to highway congestions [6]. For instance, access to live "traffic" charts would aid the inhabitants in choosing the suitable path to conserve time and effort.

The extensively utilized portable products such as "Google Maps" or "Apple Maps" precisely forecast transportation blockage for city lanes centered on the sensor information from scrutinizing tools established on roads or metropolitan highways [8]. These product contributors determine collaborations with several transport units to collect transportation statistics. The transport regulating organizations mainly establish the transport regulating tools on city highways, therefore these product contributors (e.g., "Google application programming interface") provide information on metropolitan rush-hour "traffic" clogging. Furthermore, such products additionally utilize crowdsourcing with site-established facilities [9] to enhance travel degree forecast. They do anticipate intelligent tools inside the automobile or any intelligent portable tool with the driver of the automobile to obtain live transportation information. The worry at this juncture is that the customers need intelligent gadgets to use these products and largely the services are restricted to metropolitan highways.

The travel design of metropolitan lanes is separate from that of "collector" streets. The operators of "collector" lanes comprise walkers, bikes, motorcycles, and additional automobiles; thus, the transport model is distinct from the highways. Along with metropolitan highways, the live scrutinizing of collector highways is additionally vital to enhance the movement of the whole town. The live transport jam information, alerts from transport agencies on non-recurring travel events like collisions spilled loads, VIP trips, ambulance facilities, or some additional unique road events would assist the collector lane "drivers" in their choice-making.** For example, closed properties like "universities" and "hospitals" cope with severe automobile jams during "peak hours". These properties would possess additional collector lane sections of diverse distances which would link to separate entrance/departure spots. The live transport updates of lanes that link to the departure spots would assist the driver in choosing the optimal path from his existing spot. The motorists desire to understand the clogging state of upcoming junctions to strategize and conserve their time on the highway by selecting alternative routes. The issue that develops here is in what way to deliver live highway clogging updates to motorists in the absence of intelligent devices within or outside their automobile.

The latest investigation attempts in smart transportation methods reveal that the "IoT" model would perform a significant part in travel administration by linking the basic gadgets over the net to exchange data, tracing, and observing transportation motion [10–12]. The "global positing systems", devices, investigation automobiles, and automobile to transportation interaction are limited methods to gather live highway information. The devices such as sound and magnetic sensors are price-efficient, energy-effective, and very prevalent amongst the latest automobile examining results [13, 14]. The gathered traffic data from various sources could be utilized to forecast and handle traffic jams. Most of the prevailing solutions provide live travel updates of metropolitan highways particularly via intelligent portable tools [8, 14].

5.1.2 Big Data Analytics and Its Role in Traffic Management Systems

"Big Data" is a method of collecting, regulating, and assessing the information to create expertise and disclose unseen patterns. The emergence of "Big Data" has caused troublesome deviations in numerous domains comprising "Intelligent Transport Systems (ITS)" with an extensive array of products from intelligent metropolitan design to improved automobile security. Nevertheless, methods and rules in several areas of ITS have not maintained pace with the rise of "Big Data". More precisely, the present traffic regulator methods such as "feedback loop" or standard analytical techniques are not tailored for "Big Data Analytics". The growth of the present "ITS" into an information-powered approach is predicted by numerous scientists [15].

In "Big Data" methods, the task is not any longer to gather the information but to derive useful decisions by appropriately examining them. To be precise, manipulating the accumulated information has been constantly considered by scientists and experts, however, the excessive rate, scale, and heterogeneousness of the huge flow of live information pushes the restrictions of the existing storage area, supervision, and administering capabilities. Indeed, the traditional numerical techniques are questioned (particularly with regards to bias) and are not applicable to the evolving information flows. Several of these information flows are organized in a manner that operates merely on predefined purposes and is not directly utilized for additional means. However, there are evolving unstructured information such as situation-centered information [16] from the net and community channels along with credit card operations which are vague but can be employed to better comprehend the flexibility models. Consequently, it is vital to building advanced method concepts which permit us to effectively manage huge and innovative information flows.

Although the number of experiments on "Big Data" in transportation has significantly risen, most of the installed products that aid "Big Data analytics" in "ITS" depend on unstructured design [17, 18]. They concentrate on fulfilling certain predefined objectives (extracting GPS information, forecasting traffic movement, etc.) and are difficult to apply on distinct products and information resources. This results in inflexible methods and overall constrain the acceptance of "Big Data" tools in "ITS".

The remainder of the chapter is structured as follows. Section 5.2 describes the different technologies being utilized in ITMS. The basic working of ITMS is discussed in section 5.3. Different case studies on Smart TMS are discussed in section 5.4. Section 5.5 provides an overview of the research domains that exist in this area along with the applications of ITMS. Section 5.6 illustrates the different challenges faced in building smart city applications. The summary of the survey is given in section 5.7 followed by conclusion and future scope in section 5.8.

5.2 Technologies Used in Intelligent Traffic Management Systems

ITMS is built using a combination of technologies, as stated below. The technologies must be selected based on their intended usage, location, and budget. Advances in telecommunications and information technology, as well as the availability of ultramodern and state-of-the-art microchips, radio frequency identification (RFID), and low-cost intelligent beacon-sensing technologies, are assisting in the enhancement of these systems' technical capabilities [19].

- **Communication technologies**: Radio communication using UHF and VHF is the most common wireless communication technology utilized for ITMS. Longer-range communications use infrastructure networks such as Worldwide Interoperability for Microwave Access (WiMAX), Global System for Mobile Communications (GSM), or 3G.

- **Technologies for automotive electronic systems**: In vehicles, these methods are utilized for engine supervision, ignition, wireless, laptop-on-panel, "telematics", "in-car entertainment", steering, and associated components of additional power structures. Advanced driver-assistance schemes such as track-support, speed-assistance, screen location recognition, playground-assistance, adaptive travel management, pre-crash aid, and others may be included in automotive electronic systems. Latest developments in "hardware" memory organization and "real-time operating systems" have allowed in-automobile automotive electronic systems to utilize less and another efficient computer "microprocessors". In the initial 2000s, most vehicles used simulated "microprocessor" based "operating systems", while today's "embedded" systems incorporate significantly additional advanced systems and package products, as well as AI and pervasive computation. Traffic-measurement tools to obtain raw data for travel volume, several tools are offered. To accurately determine the state of the traffic, information from various techniques can be blended in a smart fashion. The advantages of different individual methods are combined using an information-synthesis-centered methodology that uses roadside accumulated audibility, imagery, and device information. Normally, traffic measurement is done with cameras; however, innovative, quicker, and additional trustworthy techniques utilizing sensors are now accessible. These are less costly than cameras, deliver additional analysis (possibly containing all places and lanes), are easier to set up, require a smaller amount of care, and work in all climate situations [20].

- **Video automobile-recognition approach:** Transport movement monitoring and automated scene recognition utilizing video cameras are the highly prevalent types of automobile detection. Cameras are installed on constructions over or next to roads. This method of traffic sensing is non-intrusive because no components are put immediately into the highway exterior or highway-bed. Processors use audio-visual material from the cameras to examine the altering features of video imageries as automobiles go through. A specific "video detection processor" could identify traffic from up to 8 "cameras" at the same time. The method involves certain preliminary planning to acquaint the CPU with the standard environment imagery by maintaining established metrics as input, like the space among road tracks or the camera's height over the highway. Road-by-road automobile velocities, totals, and road-use measurements are among the information provided by a video vehicle-recognition method. Gaps, headway, stopped-vehicle detection, and incorrect-path automobile alerts are a few of the other outcomes that particular methods can provide [21].

- **Audio vehicle-detection system**: The audio signal can also be used to evaluate traffic density because it contains noise from tires, locomotives, engine-running noise, beeps, and air uproar sound. Using audio from a pavement microphone that collects up a variety of automobile sounds, an acoustic-motion-handling technique is utilized to evaluate the situation of transportation. When compared to other systems, such a system has outstanding accuracy [22].

- **"Sensing systems"**: These are automobile and transportation-centered linked structures. Transportation devices like in-lane indicator appliances that are connected or fixed on the highway or on roadsides remain durable. Automobile-detecting schemes install transportation-to-automobile and automobile-to-transportation electric signals for the detection of networks. These might also utilize audio-visual automated license plate-identification or automobile magnetic-sign-detection tools at periods to enhance constant examining of automobiles in vital regions. "Inductive" rounds are in a roadbed to perceive automobiles as these pass across the "loop's magnetic field". The sensors merely estimate the volume of automobiles that cross the loop, while good sensors assess the velocity, size, and category of automobiles, and the gap amongst these. Loops can be positioned in a separate road or around several roads [23].

- **Methods centered on cell phones**: Nowadays a great percentage of automobiles comprise of multiple cell phones which regularly communicate their existing data to the cell phone systems even in the absence of voice connections. As an automobile starts, the signal of the cell phone also starts. By assessing and examining web information utilizing triangulation, design matching or unit-region data, this information can be transferred into transportation stream data. The benefit of this approach is that no extra structure needs to be developed down the path since the current cell phone links remain utilized. Nevertheless, the technique can be difficult in regions wherever the identical cell phone tower serves 2/more similar roads [24].

- **Automobile re-identification**: This technique needs sets of sensors installed near the lane. Here, a distinctive license number of the automobile is identified at one site and later identified again (re-found) additional along the highway. Journey period and speed determined by comparing the period at which the automobile is identified by the sets of devices. This can be accomplished utilizing "Bluetooth, RFID" or license plates from "electronic toll collection (ETC)" "transponders" (additionally known as "toll tags") [25].

- **GPS-centered technique**: An escalating volume of automobiles remain currently fitted by in-automobile "GPS" or "satnav" ("satellite navigation") methods which possess 2-way transmission with a transport supplier. Numerous location evaluations from these automobiles are utilized to calculate automobile speeds [25].

- **Smartphone-centered checking**: "Smartphones" with devices are utilized to monitor travel velocity and intensity. "Accelerometer" information from "smartphones" utilized by automobile motorists is examined to determine the traffic speed and lane condition. Acoustic information and "GPS" labelling of "smartphones" facilitate the detection of travel density and probable traffic stops [25].

- **Bluetooth identification**: Sensing mechanisms embedded down the highway are employed to identify Bluetooth methods prevailing in automobiles. Through connecting these devices, it is probable to compute the journey period and deliver information for source and target patterns. This is an exact and economical approach to assess the journey period and get the source and target assessment. In comparison to other transport metric tools, "Bluetooth" dimensions possess the benefits of being more precise and non-invasive. This makes lesser-price connections for stable and transient locations feasible that are normally ready to set up with slight measurement needs. In this technique, calculating and additional applications are constrained by the number of "Bluetooth" methods transmitting in an automobile [25].

5.3 Basic Working of an ITMS

The main aim of ITMS is to help road authorities to increase the overall effectiveness of the transport system. This support is through all facets of highway link. It is vital to resolve transportation difficulties by mixing skills, such as automobile-steering scheme, transportation-sign-regulator scheme, flexible communication symbols, automated "number plate" identification or "speed cameras", and additional organizations that mix live statistics and response from numerous "sources", like "parking" management and info schemes, climate data, "bridge de-icing" schemes to name a few. Furthermore, analytical methods are utilized for sophisticated modelling and assessment with historic standard information. Centered on real-time and historic information, transport indicator phases are changed on a live basis, utilizing in-situ devices or investigation methods, to react to varying travel situations. A naive and straightforward scheme involves a transport-management facility, a transport data maintenance facility, a portable transmission structure, and an in-automobile station ("GPS tracer"), linked with one another through a radio transmission link ("satellites" or "telecom towers"). The in-automobile station connects through the transport power hub and the travel data facility correspondingly through the "cellular mobile communication system" ("satellites" or "telecom towers") [26].

The commuter-monitoring facility offers live information linked to travel intensity in every road to the significance system device that is inserted in the intelligent frame linked to "traffic lights". The regulator handles the information and communicates monitoring signs to the intelligent "traffic lights" for changing the "traffic lights" centered on transport intensity in every road. Information gathered by the transportation-regulator hub supports rule-applying activities by delivering live data as rapidly as a stolen automobile is perceived, supervising carbon discharge by automobiles through synchronizing to the database of contamination examination central headquarters, delivering automated and live data of any accident taking place on roads, mixing an automatic digitized scheme for detection of transportation destructions and delivering an operational "e-challan" to automobile holders, and delivering a "clear" street for smooth drive of transportation throughout "emergency" circumstances [26].

5.4 Case Studies on ITMS

This section discusses certain relevant case studies on ITMS.

5.4.1 ITMS in India

Developing nations like India have begun taking initiatives for building ITMS in big cities. For example, in Delhi, currently more than 800 million rupees is being invested for ITMS that encompasses more than 200 kilometers of roads possessing 200 plus traffic lights. In Bengaluru, 100 plus cameras have been installed at various intersections for monitoring remotely. This assists in improving the management of traffic effectively. More than 100 traffic signals have been linked to the TMS. i.e., when no vehicles transit for a period of 4 seconds, the traffic signal automatically goes red. Similarly, Mumbai has installed more than 650 plus intelligent devices across several intersections and junctions to control traffic signals in an automated way. Using this information from devices the ITMS can change the traffic signals based on dynamically varying traffic conditions. Once this system is fully implemented, it could effectively reduce the waiting period at traffic signals [27].

In another metro city of India (Chennai), approximately 100 camera sensors have been mounted in several intersections to monitor and regulate the traffic signals. It is currently operational and can identify waiting vehicles in numerous traffic junctions. In the city of Mysore, the KSRTC is preparing to develop ITMS aided by "World Bank". The implementation scope comprises of several steps of strategy, analysis, preparation, etc. for a duration of 3 years. The work comprises of 400 plus buses, 70 plus bus stops with 9 plus bus stations. The system would possess regulatory settings for managing passenger information, travel requirements, vital power base, commuter data supervision scheme, transmission sub-system, incident and disaster supervision method, and navy supervision method. The "Centre for Development of Advanced Greater Hyderabad" possesses a principal strategy for applying "ITMS" to assist optimization of transport stream, lessen ecological influence, decrease motorway collisions, and encourage effectiveness in transportation and its administration. There are additional ventures that have been planned around "India", such as the application of "ITMS" in "Indore, Pune" to name a few [28].

5.4.2 Exigency Ambulance Management System

In this effort, authors devised a system where they used technologies like "IoT", "Global Positioning System (GPS)", and "cloud" to supervise the "ambulance" sites and choice of paths. A pre-informing technique was also designed to keep the related stakeholders informed, via a Hop-Siren system that could be installed at the junctions. The scheme was intended to notify the Transport Workers regarding an oncoming "ambulance" well ahead for the efficient management of transport and for creating a green passageway to the "ambulance". A "Web/Mobile app" is employed as the "user interface" to reserve an "ambulance" that additionally listed adjacent dedicated clinics. After a hospital was selected, either by the customer or by the "ambulance" staffs, the optimal path from the position of occurrence to the hospital was delivered through the "Google maps APIs". The transport management workers interface kept the agencies abreast about the latest ambulance's path [29].

The designed structure is comprised of three units that are discussed as under:

1) *"Ambulance Unit (AU)"*: The "AU", to be inserted in the "ambulance", consists of a "GPS system". The "GPS receiver" consistently gets the "GPS coordinates" of the "ambulance" and utilizing the "Google Map APIs" it is communicated further to a central server. The location and routing of the "ambulance" to the closest hospital of option is operated by the central server. The implemented APIs offer precise live path arrangements to the "ambulance" operator.

2) *"Hop Sirens"* / *"Junction Unit"*: The "Hop Sirens" are triple light markers integrated with an alarm device to point out the position of the oncoming ambulance. These components are positioned at the transport junctions. The administered information from the "GPS" tools in the "AU" is acquired by the "transceivers" in the hop alarms and therefore they modify lights aptly. The "3-light" system is described below:

i) Initially "light" flashes with a sparingly recorded alarm sound. This suggests that an ambulance is slated to go across the junction and that the concerned persons are to be mindful and make sure that they are not an obstacle to the "ambulance" flow.

ii) Another "light" flickers with a moderate planned alarm bleep. This implies that the "ambulance" is near the junction and is only some hops or moments away and that the obstructions require to be discharged and guaranteed and no further piling happens.

iii) 3rd "light" flashes with a relatively noisy alarm bleep. This signifies that the "ambulance" is inside the junction area and that the transport flow requires to be retained transparent adequate for its journey.

Built on the information conveyed through the "hop sirens", the transport staff needs to participate in emptying the roads for the "ambulance". Additionally, this data can be utilized to spontaneously influence the road "traffic" indicators to make sure all the necessary roads are emptied for the "ambulance" traffic, guaranteeing not to cause hassle to the regular travellers. This inconvenience to the community can be prevented by slowly adjusting the "green-light" for the broad roads in a non-perceptible period adjustment in a "round-robin" fashion [29].

3) *"User Application"*: There are various "user interfaces" as explained below:

i) Community user: The client obtains an "interface" to call an "ambulance" and this could be accomplished via the "application" or the "Emergency Numbers" ("108").

ii) Transport staff: The transport staffs obtain particulars of the "Ambulance" sites via the "maps interface".

iii) "Ambulance Management Team": The "ambulance interface" provides the driver with the path specifics.

iv) "Hospital unit": The "hospital unit" is designed to safeguard the medical personnel by informing them early so that they can be trained to deliver health care for the patient in a timely manner [29].

Figure 5.1 represents the devised system's design.

After the "ambulance" aid is needed, the app could be utilized precisely to call the "ambulance" or call the crisis number ("108"), for entering the specifics in the "app". "Cloud" will appropriately choose the "ambulance" that could achieve the "Patient" site by taking into account travel conditions. When the "patient" specifics are gathered, the "cloud system" chooses the appropriate clinic, unless explicitly asserted by the consumer. The "Google Maps" are utilized for transmitting. The appropriate specifics are additionally distributed through the clinic "servers" for data and to organize themselves for the approaching serious incidents. The transport patrol administration front is also notified regarding the "ambulance" travel particulars for improved aid. Later utilizing the "map interface", each of the "hop sirens" and the corresponding transport patrol is informed regarding the travel specifics. The 1st lights in the hop sirens are turned ON. The nearest markers to the "ambulance" are then suitably controlled to lessen the "green" period on the "non-distress" roads discreetly with respect to seconds so that the "green" period on the road is enhanced. This would consequently enable the quicker passage of the "ambulance" all through, decreasing the travel time and protecting natural life.

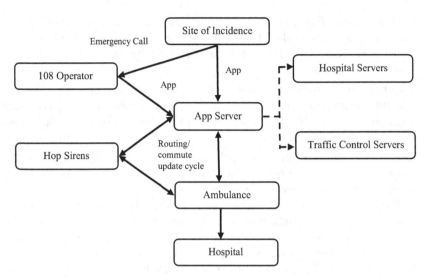

FIGURE 5.1
System design.

The researchers devised an end-to-end comprehensive system to guarantee decreased reaction speeds of "ambulances" in case of crises. This computerization method offers a resolution starting after the ambulance request till the patient is taken to the health center. The approach connects via "GPS", "Cloud" tools, and "IoT" and consequently applies a viable and suitable resolution to the practical problems in Metros. The recommended system was assessed by executing the proof of concept utilizing the "Arduino microcontroller". The signal control was extremely swift and the idea could be applied and assessed on practical scenarios. The plan of the effort was to enable quicker ambulance movement by lowering traffic jams on the anticipated lanes. The authors observed that this could decrease the total travel time taken by emergency vehicles in the traffic [29].

5.4.3 Case Study-3

In [30], the authors designed an ITMS. The strategic goals of the effort are reviewed below:

- **Traffic Monitoring (TM)**: It is one of the crucial elements of a smart metropolis. TM permits the regional agencies to examine the movement of traffic-related to a specific region, road, or motorway. It facilitates in maintaining the path of the traffic inflow from other adjacent towns through particular days or times of the year. Historic information of transportation supervising could be very valuable in smart city development.

- **Pollution Prevention**: Increasing pollution levels create a danger to the ecosystem alongside the harmful effects on social wellbeing and fitness. The magnitude of air and sound "pollution" is "directly" proportional to the volume of transport bottleneck in a municipality. Lengthy lines of automobiles cause excessive pollution causing global warming, reduced rainfall, breathing difficulties, etc.

- **Path Optimization**: In modern times, it is noticed that the shortest path may not be optimal with respect to the overall journey time, fuel expenditure, and normal delaying period. In such situations, an optimal path is a top alternative for commuting since it considers issues like transportation bottleneck, distance covered, overall commute period, and fuel intake. An optimal path includes a trade-off among all these factors and saves the time and fuel for the commuter.

- **Green Corridor**: Recently, the notion of a green corridor has risen. It is a passageway which is a path from a source to the target including numerous traffic signals every one of which possesses a green sign. The "green corridor" is utilized to accommodate the "emergency" automobiles by permitting them to attain their preferred target minus some delay time and at full pace.

- **Identification of collisions**: The congested situation of present-day lanes has increased the volume of accidents. Collision discovery is a vital component of a transport supervision scheme since it not only notifies the health personnel to show up to the accident site but additionally impacts the traffic movement and overcrowding levels of a specific area.

- **Congestion**: Avoidance of traffic clogs and decrease in regular waiting period are the two most crucial features of an effective TMS.

- **Automobile Tracing**: It facilitates the regional administration in maintaining the trace of automobiles with respect to the regions they are moving, commute time, velocity, locations visited, and automobile category. All these factors turn out to be productive when it comes to preserving law and order.

Figure 5.2 shows the layered architecture

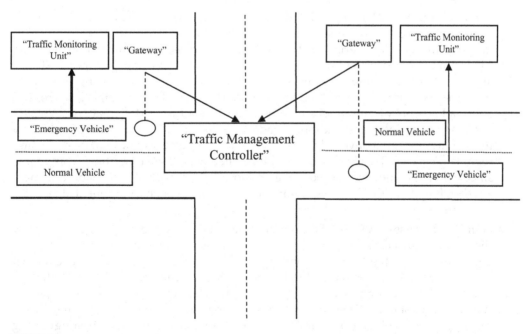

FIGURE 5.2
ITMS architecture.

i. **"Traffic Management Controller (TMC)":** The objective of the regulator is to regulate and administer the whole system [31]. It arranges the tasks of additional application components and units inside the structure. The controller is installed at the Cloud end and possesses complete data concerning each automobile, "traffic signal", entrances, On-Highway Devices, and Transportation Supervision Module. All this data is deposited and handled by the regulator to produce improved paths among the particular source and target. It creates a "one-to-one" link by the "middleware" and distributes orders through it. It produces forecast information regarding the amounts of transport clogging at differing periods. The "TMC" employs a flooding algorithm based on the hop counter for transmitting reports concerning a mishap, a shift of paths, highway developing events, and unfavorable climatic effects. The TMS is also responsible for the generation of a "green corridor" when "emergency" vehicles are nearby.

ii. **Gateways:** The entire data which was detected and gathered through the "on-road sensors" are transferred to the "gateways" [32]. They behave as a regular point of communication in which different types of data emanating from diverse kinds of sensors are gathered. The gateways employ a greedy-based information gathering procedure for accumulating information from a variety of information resources. It does universal addressing of Automobile Nodes (V) by using "IPv4 addresses" [9]. Every entryway is allocated a coverage zone, in which every on-road device and V are provided an IP address consequently enabling effective detection of entities inside that region. Each "gateway" is allotted an extra region to improve the quality of automobile recognition. The "gateway" additionally maintains the trail of its nearby entryways alongside the automobile nodes commuting in its region. Lastly, the entryway communicates all types of amorphous data to the TMC.

iii. **"Traffic Monitoring Unit (TMU)":** It functions as an intermediate connection among "On-Road Sensors" and "Gateways". The objective of inserting a "TMU" was to improve the reaction period of the system since direct communication with the "TMC" may increase the expense. It offers a transmission linkage among "TMC" along with the remainder of the system. It performs local managing and space resources to improve the productivity of the system [33]. The data emerging from an "on-road sensor" or automobile node is resolved via the "TMU" that consequently notifies the Controller and additional gadgets on the grid. The directions provided by the "Controller" are conveyed via the "TMU" to the relevant automobile nodes and regional agencies. The TMU can also be deemed as a "fog computing" component since it dwells at the boundary of the grid allowing it to be easily accessible and manageable. It updates the TMC regarding the data of each system component at periodic intervals.

iv. **"On-Road Sensors (ORS)":** Sensors are vital since they are employed to identify the event occurrences, neighboring circumstances and communicate the gathered data. The task of the ORS is to observe and recognize occurrences or events that occur on roads. Each ORS could be classified using three factors viz., sensor category, approach, and sensing factors. Sensor category characterizes the sensor type i.e., whether it is a standardized/diverse sensor or is a 1-D or a multidimensional sensor. The approach involved differs in the way a sensor collects data [32]. Distinguishing factors signify the type of parameter which it can detect. It may just sense one factor such as body heat, or several factors similar to the instance of an "ECG". Every "sensor node" is offered an "IP address" that aids in its distinctive recognition. It connects all its sensor information to its ensuing entryway. The functionalities offered by ORS are automobile tally, presence, speed, categorization, and Minimal Bandwidth Intake.

v. **Automobile Node:** It is the automobile for which a whole transport structure is built to offer an uncomplicated and comfortable commuting experience. It is also a mobile sensory connection that receives and transmits data while commuting. Every automobile node is offered an "IP address" that aids in its distinctive detection. Each "sensor node" transmits all its device information to its consequent "gateway". Each transport automobile possesses an "LED display" fitted which notifies the navigator regarding the optimal path and the continuously altering amounts of transportation. Each communication or alert like the collision alarm or avoidance of access in a specific region from the "TMC" could be observed on the "LED display".

5.4.4 Case Study-4

In [14], the researchers devised an "IoT" centered method to gather, administer, and gather live transport information for such a situation. The aim is to offer live transport updates on vehicle clogging and bizarre automobile events via pavement communication components and thus enhance movement. These initial alert communications can assist people to conserve their time, particularly in peak times. Additionally, the system transmits traffic updates from governmental agencies. A model is executed to assess its viability. The outcomes of the trials demonstrate high precision in automobile identification and a minimal comparative inaccuracy in highway use assessment. The devised method is illustrated in Figure 5.3, which comprises elements mounted at the pavement and a "cloud-based central server".

The pavement system comprises "sensors" and communication panels. The sensors and panels will be fitted amid 2 highway division junctions. The central server comprises information space, cloud facilities, and interfaces. The modules can transmit with each other utilizing Wi-Fi. An "IoT" centered design mainly comprises detecting, link, maintenance,

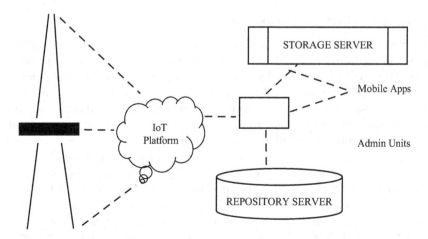

FIGURE 5.3
System communication model.

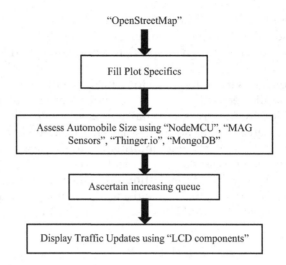

FIGURE 5.4
System development activities.

and "application" layers [34]. The detecting layer secures information from the devices, the link-layer transmits the gathered information from gadgets to the maintenance layer, the "service layer" regulates the gadgets and examines the gathered information, and ultimately, the "application layer" shows the "user interface".

The four key product enhancement actions are as follows: (i) populate the geographic chart with the specifics of an allotted site, (ii) identify automobile and assess automobile size, (iii) ascertain expanding queue, and (iv) exhibit transport updates. The method elements consist of (i) Geographic chart, (ii) "Sensors", (iii) "Microcontroller", (iv) "IoT" platform, (v) Databank, and (vi) Automated demonstration components. The events, the package, and "hardware" elements linked through every action are provided in Figure 5.4.

The scheme enhancement measures are outlined below:

5.4.4.1 Geographic Chart Information Handling

The geographic chart gives the road sector data, junctions, and lanes. These plots are handled to pile the road data to the databank and mine the communication panel sites. The client-created chart could be utilized to obtain the communication panel site [35]. The highway intersections which possess additional linked highway sections form the superior sites to exhibit transport-alert communications. The communication panel sites are chosen using their coverage to boost transmission visibility. This is deemed as an extension issue since the goal is to boost the visibility of the message. The notion of poster marketing may be employed at this juncture to increase the intensity of communication coverage [36]. Additionally, the main parking spaces in a sealed property could be chosen to get the communications to the highest volume of people. The number of linked roads is one factor that determines the intensity of information coverage. Likewise, the previous forms of transportation intensity could also be chosen while ascertaining the message component site. The "OSM" chart of a site could be taken through the "OpenStreetMap" website. The "OSM file" supports an "XML" structure, and it has three main components: nodes, approaches, and relations. There is a diverse label to recognize every kind of road. The road key may possess distinct values like a highway, chest, primary, domestic, etc. Likewise, the connection highways may be detected. The intersections key could be utilized jointly with motorways, and specific kinds of intersections are junctions, spherical, filter, and jug handle. These keys assist to mine the appropriate data on a road. The procedure starts with geographic chart system transformation and databank packing. The subsequent action is to recognize the communication panel sites cantered on earlier transport intensity on junctions and the volume of linked lanes.

5.4.4.2 Automobile Recognition and Physical Size Assessment

The live automobile information sensed by the "magnetic" devices is employed to identify automobiles and assess their size. The estimated automobile velocity is a factor that is mainly utilized to ascertain the size of the automobile [37, 38]. "Magnetic sensors" identify the disruption in the globe's "magnetic field" triggered by mobile automobiles and computed as automobile "magnetic" size [39]. The "vehicle magnetic length (VML)" is employed to assess automobile size [40, 41]. The automobile speed assessment utilizing two magnetic sensors has exhibited superior precision when compared to a solo sensor. This study employed the notions implemented in [42], i.e., the "sensors" were fitted with a gap "d" and no path switching occurs. The speed was determined from the journey period among 2 parallelly positioned magnetic device points "A" and "B" with a gap "d". The arrival ("arr") and exit ("dep") times were loaded once the strength reached above and fell below the limit, correspondingly.

5.4.4.3 Lane Occupancy and Increasing Queues

The transport bottleneck methods are mainly centered on factors like "speed", "time" and "delay", dependability, maintenance, space, to name a few. The highway area use is such a metric to ascertain the increasing congestion. Figures 5.5 and 5.6 demonstrate the flow diagram on how to ascertain the highway bottleneck utilizing occupancy metrics.

Figure 5.7 demonstrates in what way the road use metric is computed.

The "VPL" was assessed through the "sensors" after the automobile crossed the "sensor" nodules, the real size was included in the lane tenancy metric and deducts the size once the automobile leaves the departure spots as shown in Figure 5.6. After "sensor C" identifies an automobile, "sensor D" assessed the real size and transmitted it to "sensor B". The "microcontroller" linked through "sensor B" retained the "occupancy" metric and transmits live transport updates. The lane "occupancy" metric was assessed using the automobile size at the entry and departure of the lane section. The procedure in Figure 5.5 was executed as "firmware" for the "sensor B" in Figure 5.6. The "sensor D" merely approximates the real size and transmits it to node A, which functions as the server and delivers the information to the "display" module. It also loads the information to the "IoT platform" wirelessly.

5.4.4.4 "Display" Alert Communications

The transportation alert communication can be of 2 kinds: (i) live updates on transport intensity, (ii) communications on rare highway events by agencies. These updates are

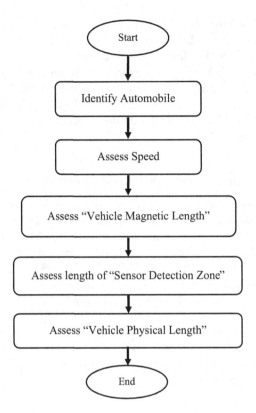

FIGURE 5.5
Flow chart for road congestion detection using occupancy metric.

accessible by motorists while operating via various forms like intelligent portable products, transistors, TVs, to name a few. The alternative approach was to employ pavement communication panels at major junctions. These modules will reach most people and assist them to take alternative paths.

5.5 Research and Applications of ITMS

Some of the real-world applications of ITMS are presented below:

- **Automated road enforcement:**

  ITMS offers an efficient commuter-enforcement scheme to identify and find automobiles violating speed restrictions or other legal constraints. It routinely books the lawbreakers using their license plates. Traffic tickets are subsequently mailed to the vehicle owners. The scheme involves [43]:

  - "Speed cameras" to detect "over-speeding" automobiles. Several techniques additionally utilize "radar loops" hidden in every travel path to detect automobile speed.

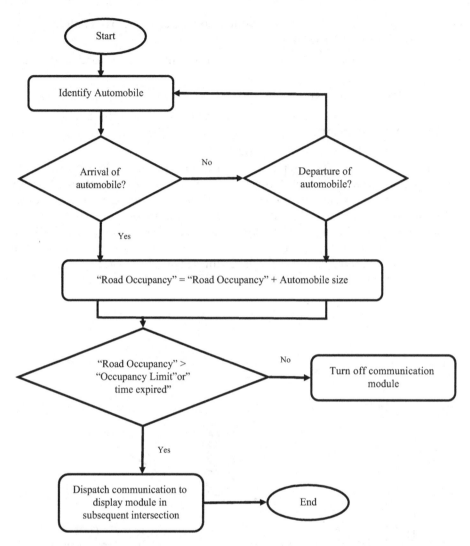

FIGURE 5.6
Detailed flow chart for road congestion detection using occupancy metric.

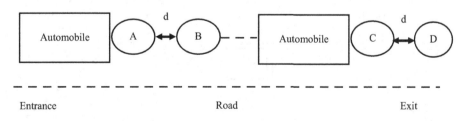

FIGURE 5.7
Scenario to assess highway occupancy.

- "Traffic light cameras" to identify automobiles that go across a halt line while a "red traffic light" is displaying.
- "Bus lane cameras" to detect automobiles moving in roads booked for "buses".
- Dual colorless stripe cameras that detect automobiles traversing these areas.
- "High-occupancy vehicle (HOV)" road "cameras" to detect automobiles breaching "HOV" constraints.
- **"Dynamic traffic light sequencing" (DTLS):**
  Utilizing "RFID" for DTLS bypasses or prevents challenges that typically occur with systems using "image-processing" and "beam-disruption" methods. "RFID technology" with suitable system and databank are employed in a multi-automobile, multi-road, and multi-lane intersection region to deliver an effective "time-management" program. The "dynamic" sequence procedure adapts itself to handling extreme circumstances. The system can imitate the decision of a transport patrol sergeant on the job, by considering the number of automobiles in every column and forwarding proprieties. Figure 5.8 shows the "Dynamic Traffic Light Sequencing" [44].

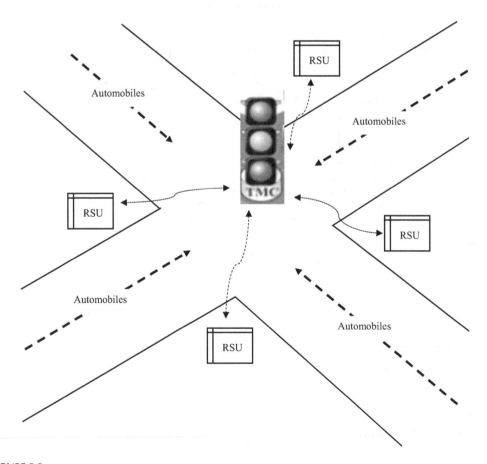

FIGURE 5.8
"Dynamic traffic light sequencing".

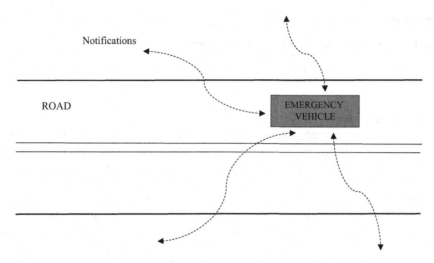

FIGURE 5.9
"Emergency vehicle notification" system.

- **"Emergency vehicle notification system"**
An in-automobile "e-call" is created physically by the automobile occupier(s), or mechanically through the initiation of "in-vehicle" devices following a crash. The "e-call" transmits the voice and information promptly to the closest trauma center (generally a community protection responding center). The voice call permits the occupant(s) of the automobile to interconnect with the qualified e-call worker and concurrently the smallest set of information comprising of automobile ID, stats regarding the event, period of incidence, exact site, and course in which the automobile was going are directed to the e-call worker for taking required steps. In several advanced nations, e-call is a basic alternative for every authorized automobile [45]. Figure 5.9 shows the Emergency vehicle notification system.

- **Highway transport collaborative system:**
A collaborative system in highway transport involves highway operatives, transportation, automobiles, their handlers, and additional highway operators collaborating to provide the most effective, reliable, protected, and relaxed voyage. The system is centered on transmission collaboration on the highway. Information accessible from automobiles is gathered and communicated to a server for processing. This information can be utilized to identify occurrences like rain (windshield wiper movement) and blockage (repeated decelerating movements). The server procedures a steering suggestion devoted to a solo or a particular band of motorists and communicates it "wirelessly" towards the automobile(s). A collaborative approach aims to utilize and develop interaction and sensor structures to assess highway security [46].

- **Altering "speed" restrictions centered on road blockage**
Investigation with varying speed restrictions that alter with road blockage and additional issues is being researched. Normally, these speed restrictions decline in poor circumstances, instead of improving in decent situations. One instance is UK's M25 highway, which goes around London. Flexible speed restrictions merged with automatic implementation provide savings in travel periods, easier-running road traffic, and a drop in the number of crashes [46].

5.6 Challenges in Smart City Applications

Certain challenges faced in developing applications for a smart city are highlighted below [47]:

- **Infrastructure:**
  Smart cities make use of sensing devices to capture and examine data to enhance citizens' quality of life. Sensors gather data on everything from traffic congestion to crime rates to air quality. The equipment and upkeep of these devices necessitate a complex and costly infrastructure. They need to be powered using hard-wired/ solar-powered/battery-powered. In case of an energy outage, a combination of all these three needs to be utilized. Major cities are already dealing with the challenges of updating decades-old infrastructure including subway cabling, steam pipelines, and transport channels, as well as adding fast internet. Although the availability of broadband wireless service is improving, there are still some locations in big cities where connectivity is restricted.

  New infrastructure projects have a finite amount of funding, and approval processes might take years. People who live in these cities face temporary – but nevertheless frustrating – challenges because of the installation of new sensors and other enhancements. Developers can make it easier to install and use smart technology by thinking about these issues early in the development process. Developers and tech businesses can speed up the process of making the cities smarter by adopting easy-to-install technology by starting with the end in mind — that is, the full implementation of the solution.

  For instance, the City of Oshawa has entered Infrastructure Canada's Smart Cities Challenge in collaboration with key stakeholders, to produce smart city solutions that draw attention to local issues. The major purpose is to cooperate with people, businesses, academic and civic bodies to discover shared problems and build new projects that address their most important challenges using data and connected technology.

- **Hacking and Protection:**
  As the use of IoT and sensor technology grows, so does the level of security danger. This raises the question of whether technology is truly "smart" if hackers can get into it and shut down an entire city. Everyone is a little more anxious and skeptical about technology and security after recent discussions about cyber-terror risks to vulnerable and ageing power grids. Smart cities are allocating more funds and resources to security, while tech companies are developing solutions with new built-in safeguards against hacking and cybercrime.

- **Concerns regarding confidentiality:**
  There is a balance in any major city between the quality of life and violation of privacy. While everyone wants to live in a more convenient, quiet, and healthier environment, no one wants to feel observed all the time. Cameras on every corner may reduce crime, but they can also instill fear and paranoia in law-abiding residents. Another genuine issue is the amount of data collected from all the smart sensors that residents come into contact with on a daily basis. Developers can assist citizens of smart cities to feel more at ease by including openness and education in their solutions. Companies may build confidence from the consumers their solutions are supposed to benefit by

creating with the community in mind and understanding how they might respond to new technologies. Local government authorities and community boards must, of course, be involved in the implementation and instructional components.

- **Educating and Involving the Public:**
  To fully exist and develop, a Smart City requires "smart" residents that are engaged and actively use new technologies. Any new city-wide technology initiative must include community education as part of the implementation process. This can be accomplished through a series of in-person town hall meetings and voter registration email campaigns, as well as an online education platform that keeps citizens informed and involved. When a community feels like it has a say in the big issues that affect daily life, and when it is informed in a clear and thoughtful way, it is more likely to use technology and urge others to do so as well. This is crucial to the success of a Smart City.

- **Being Socially Inclusive:**

  In a busy metropolis, smart transit initiatives that provide riders with real-time updates are a terrific concept. But there are also individuals who cannot afford to take public transportation or Uber. Older people may also be present who do not have access to mobile devices or apps. Hence, it is critical that Smart City planning considers the needs of all people other than the wealthy and technologically advanced. Rather than further dividing individuals based on income or education levels, technology should always be working to bring them together.

  Considering these individuals, as well as the other challenges described in this section, can help to ensure that a solution is successful outside the IT world.

5.7 Summary

In this survey, IoT in ITMS, Big Data Analytics and its role in ITMS, technologies used in ITMS, working of ITMS, four case studies in ITMS, along the research areas in ITMS were discussed. IoT in ITMS is being employed to collect live traffic updates. The gathered traffic data from various sources could be utilized to forecast and handle traffic jams. Most of the prevailing solutions provide live travel updates of metropolitan roads especially through intelligent portable devices. "Big Data Analytics" needs to be extensively employed in ITMS to bring about breakthroughs. It is found that they focus on only specific products and is not applicable to other products. Some of the technologies used in ITMS are communication technologies like WiMAX, GSM, technologies for automotive electronic systems, video automobile recognition, sensors, smartphones, GPS, and Bluetooth. The primary objective of ITMS is to assist highway organizations to improve the general efficiency of the transport system. This aid is across all aspects of the road link.

Four case studies were discussed. The first case study was about ITMS in India. Currently, ITMS is being explored in major cities of India like Bangalore Chennai, and Pune. The second case study discussed the "Exigency Ambulance Management System" developed by [29]. The system is comprised of three modules viz. AU, hop sirens, and user interface. The objective of the system was to free the path of the ambulance. In the third case study, authors [30] designed an ITMS for monitoring traffic, preventing pollution, identifying potential collisions, optimization of paths, and avoidance of traffic congestion to name a few. In the fourth case study, the authors [14] designed a technique centered on

"IoT" for collecting and administering real-time data. The objective was to deliver real-time traffic updates on traffic jams to avoid traffic congestions. In this study, certain real-world applications of ITMS for detecting speed violations, dynamically varying traffic lights, notification of emergency vehicles to clear the traffic, and adjusting vehicle speed based on congestion were discussed. It is observed that these applications were highly relevant and significant in the current real-world scenarios.

5.8 Conclusion and Future Scope

ITMS is a valuable invention that offers an effective answer to the prevailing structure of insufficient TMS and its associated components (such as toll payment, challan owing to over speeding, smog inspection facility, "VIP/ambulance/fire" bid route sanction, automobile robbery, etc.). All these concerns are effectively handled by ITMS to deal with the nuisance of the present traffic crisis. The success of the applied ITMS needs solid support with regards to plan, law, management, appropriate segment management, and collective act. ITMS may not succeed to provide the anticipated advantages because of the failure of functional components. These methods were applied with good success on the quicker lanes of technologically advanced Nations such as the US, Japan, and the UK. Furthermore, several scientists are putting their effort to convert the transportation structures into more sophisticated systems. This chapter provides extensive information about ITMS, techniques to build ITMS, and some related case studies. The different technologies being utilized in ITMS along with case studies on Smart TMS were discussed. An overview of the research domains that exist in this area along with the applications of ITMS was also provided.

In the future, gradually but steadily all transportation and swift transportation procedures will go on to implement a smoother approach for "ticketing" and supervision of the techniques to replace the outdated tools and procedures. "Smart cards", "token" tourism, "comfort", "ease", "speed", "in-transit" entertainment, clean station buildings, clean and green energy to power those systems will soon be a reality when the authorities and the residents begin to collaborate.

References

1. Purnendu S.M. Tripathi, Ambuj Kumar, and Ashok Chandra, An overview of Intelligent Transport System (ITS) and its applications, *Journal of Mobile Multimedia*, 17(2021): (1–3). https://doi.org/10.13052/jmm1550-4646.17134.
2. L. Broccardo, F. Culasso, and S.G. Mauro, Smart city governance: Exploring the institutional work of multiple actors towards collaboration, *International Journal of Public Sector Management*, 32(4) (2019): 367e387. https://doi.org/10.1108/IJPSM-05-2018-0126.
3. N. Buch, S.A. Velastin, and J. Orwell, A review of computer vision techniques for the analysis of urban traffic, *IEEE Transactions on Intelligent Transportation Systems*, 12(3) (2011): 920e939. https://doi.org/10.1109/TITS.2011.2119372.
4. A.L. Bustamante, M.A. Patricio, and J.M. Molina, Thinger.io: An open-source platform for deploying data fusion applications in IoT environments, *Sensors* 19(5) (2019). https://doi.org/10.3390/s19051044.

5. A. Camero and E. Alba, Smart city and information technology: A review, *Cities* 93(May) (2019): 84e94. https://doi.org/10.1016/j.cities.2019.04.014.

6. W. Castelnovo, G. Misuraca, and A. Savoldelli, Smart cities governance: the need for a holistic approach to assessing urban participatory policy making, *Social Science Computer Review*, 34(6) (2016): 724e739. https://doi.org/10.1177/0894439315611103.

7. W.H. Lee and C.Y. Chiu, Design and implementation of a smart traffic signal control system for smart city applications, *Sensors*, 20(2) (2020). https://doi.org/10.3390/s20020508.

8. B. Brindle, How does Google maps predict traffic? 2020. https://electronics.howstuffworks.com/how-does-google-maps-predict-traffic.htm.

9. Z.A. Almusaylim and N.Z. Jhanjhi, Comprehensive review: Privacy protection of user in location-aware services of mobile cloud computing, *Wireless Personal Communications*, 111(1) (2020): 541e564.

10. Z. Li, R. Al Hassan, M. Shahidehpour, S. Bahramirad, and A. Khodaei, A hierarchical framework for intelligent traffic management in smart cities, *IEEE Transactions on Smart Grid*, 10(1) (2019): 691e701. https://doi.org/10.1109/TSG.2017.2750542.

11. Y. Qian, D. Wu, W. Bao, and P. Lorenz, The internet of things for smart cities: Technologies and applications, *IEEE Network*, 33(2) (2019): 4e5. https://doi.org/10.1109/MNET.2019.867 5165.

12. J. Qiu, L. Du, D. Zhang, S. Su, and Z. Tian, Nei-TTE: Intelligent traffic time estimation based on fine grained time derivation of road segments for smart city, *IEEE Transactions on Industrial Informatics*, 16(4) (2019): 2659e2666.

13. W. Balid, *Fully Autonomous Self-Powered Intelligent Wireless Sensor for Real-Time Traffic Surveillance in Smart Cities*, University of Oklahoma, 2016.

14. Mohammed Sarrab, Supriya Pulparambil, and Medhat Awadalla, Development of an IoT based real-time traffic monitoring system for city governance. *Global Transitions*, 2 (2020): 230–245. https://doi.org/10.1016/j.glt.2020.09.004.

15. S. Amini, I. Gerostathopoulos, and C. Prehofer, "Big data analytics architecture for real-time traffic control," in *2017 5th IEEE International Conference on Models and Technologies for Intelligent Transportation Systems (MT-ITS)*, 2017, pp. 710–715. https://doi.org/10.1109/MTITS.2017.8005605.

16. E. Chaniotakis and C. Antoniou, "Use of geotagged social media in Urban settings: Empirical evidence on its potential from Twitter," in *2015 IEEE 18th International Conference on Intelligent Transportation Systems*, 2015.

17. D. Xia, B. Wang, H. Li, Y. Li, and Z. Zhang, A distributed spatial temporal weighted model on MapReduce for short-term traffic flow forecasting, *Neurocomputing*, 179 (2016): 246–263.

18. J. Yu, F. Jiang, and T. Zhu, "RTIC-C: A big data system for massive traffic information mining," in *Cloud Computing and Big Data (CloudCom-Asia), 2013 International Conference on*, 2013, pp. 395–402.

19. Riccardo Mangiaracina, Alessandro Perego, Giulio Salvadori, and Angela Tumino, A comprehensive view of intelligent transport systems for urban smart mobility. *International Journal of Logistics Research and Applications*, 20 (2016): 1–14. https://doi.org/10.1080/13675567.2016.124120.

20. Samir A. Elsagheer Mohamed and Khaled A. AlShalfan, Intelligent traffic management system based on the Internet of Vehicles (IoV), *Journal of Advanced Transportation*, (2021), Article ID 4037533, 23 pages, https://doi.org/10.1155/2021/4037533.

21. S. Latif, H. Afzaal, and N.A. Zafar, "Intelligent traffic monitoring and guidance system for smart city," in *2018 International Conference on Computing, Mathematics and Engineering Technologies (iCoMET)*, 2018, pp. 1–6, https://doi.org/10.1109/ICOMET.2018.8346327.

22. Dipak Gade and Sreeramana Aithal, Smart city waste management through ICT and IoT driven solution, *International Journal of Applied Engineering and Management Letters (IJAEML)*, 5 (2021): 51–65. https://doi.org/10.5281/zenodo.4739109.

23. Rahul Kumar, and Kunal Gupta. ITMS (Intelligent Traffic Management System). In: Pant, M., Deep, K., Bansal, J., Nagar, A., Das, K. (eds.) *Proceedings of Fifth International Conference on Soft Computing for Problem Solving. Advances in Intelligent Systems and Computing*, vol. 436, 2016, Springer, Singapore. https://doi.org/10.1007/978-981-10-0448-3_.

24. A. Garg, "A novel approach to improve intelligent traffic management system (ITMS)," in *2015 International Conference on Green Computing and Internet of Things (ICGCIoT)*, 2015, pp. 679–682, https://doi.org/10.1109/ICGCIoT.2015.7380549.

25. A.S. Putra and H.L.H.S. Warnars, "Intelligent Traffic Monitoring System (ITMS) for Smart City based on IoT monitoring," in *2018 Indonesian Association for Pattern Recognition International Conference (INAPR)*, 2018, pp. 161–165, https://doi.org/10.1109/INAPR.2018.8626855.

26. L.C. Bento, R. Parafita, S. Santos, and U. Nunes, "Intelligent traffic management at intersections: Legacy mode for vehicles not equipped with V2V and V2I communications," in *16th International IEEE Conference on Intelligent Transportation Systems (ITSC 2013)*, 2013, pp. 726–731, https://doi.org/10.1109/ITSC.2013.6728317.

27. Hrishikesh Ugale, Pushpak Patil, and Shubham Chauhan, "Design of intelligent transportation system for Smart City," in *International Conference on IoT and its Applications (ICIA 2020)*, 2020.

28. Dinh Dung Nguyen, József Rohács, Dániel Rohács, and Anita Boros, Intelligent total transportation management system for future smart cities. *Applied Sciences*, 10(24) 2020: 8933.

29. Nashma, A. J., Sanjay, H. A., Deepthi, H., and Meghana, S. R., "Intelligent control system for the seamless transit of Emergency Vehicles," in *2020 5th International Conference on Communication and Electronics Systems (ICCES)*, 2020, June, pp. 1–5, IEEE. https://doi.org/10.1109/ICCES48766.2020.9137894.

30. Abhirup Khanna, Rohit Goyal, Manju Verma, and Deepika Joshi, *Intelligent Traffic Management System for Smart Cities: First International Conference, FTNCT 2018*, 2019, Solan, India, February 9–10, 2018, Revised Selected Papers. https://doi.org/10.1007/978-981-13-3804-5_12.

31. V. Fore, A. Khanna, R. Tomar, and A. Mishra, "Intelligent supply chain management system," in *Advances in Computing and Communication Engineering (ICACCE), 2016 International Conference on*, 2016, November, pp. 296–302. IEEE.

32. A. Khanna and R. Anand, "IoT based smart parking system," in *Internet of Things and Applications (IOTA), International Conference on*, 2016, January, pp. 266–270. IEEE.

33. D. Kyriazis, T. Varvarigou, D. White, A. Rossi, and J. Cooper, "Sustainable smart city IoT applications: Heat and electricity management & Eco-conscious cruise control for public transportation," In *World of Wireless, Mobile and Multimedia Networks (WoWMoM), 2013 IEEE 14th International Symposium and Workshops on a*, 2013, June, pp. 1–5. IEEE.

34. P.P. Ray, A survey on Internet of Things architectures, *Journal of King Saud University - Computer and Information Sciences*, 30(3) (2018): 291e319. https://doi.org/10.1016/j.jksuci.2016.10.003.

35. M. Haklay and P. Weber, OpenStreet map: User-generated street maps, *IEEE Pervasive Computing*, 7(4) (2008): 12e18. https://doi.org/10.1109/MPRV.2008.80.

36. L. Wang, Z. Yu, D. Yang, H. Ma, and H. Sheng, Efficiently targeted billboard advertising using crowdsensing vehicle trajectory data, *IEEE Transactions on Industrial Informatics*, 16(2) (2019). https://doi.org/10.1109/tii.2019.2891258.

37. W. Balid, H. Tafish, and H.H. Refai, Intelligent vehicle counting and classification sensor for real-time traffic surveillance, *IEEE Transactions on Intelligent Transportation Systems*, 19(6) (2018): 1784e1794. https://doi.org/10.1109/TITS.2017.2741507.

38. B. Yang and Y. Lei, Vehicle detection and classification for low-speed congested traffic with anisotropic magneto resistive sensor, *IEEE Sensors Journal*, 15(2) (2014): 1132e1138. https://doi.org/10.1109/jsen.2014.2359014.

39. S. Cheung, S. Coleri, B. Dundar, S. Ganesh, C.-W. Tan, and P. Varaiya, Traffic measurement and vehicle classification with single magnetic sensor, *Transportation Research Record: Journal of the Transportation Research Board*, 1917(1) (2005): 173–181. https://doi.org/10.3141/1917-19.

40. S. Kaewkamnerd, J. Chinrungrueng, R. Pongthornseri, and S. Dumnin, "Vehicle classification based on magnetic sensor signal," in *2010 IEEE International Conference on Information and Automation, ICIA 2010*, 2010, pp. 935e939. https://doi.org/10.1109/ICINFA.2010.5512140.

41. S. Taghvaeeyan and R. Rajamani, Portable roadside sensors for vehicle counting, classification, and speed measurement, *IEEE Transactions on Intelligent Transportation Systems*, 15(1) (2014): 73e83. https://doi.org/10.1109/TITS.2013.2273876.

42. W. Balid and H.H. Refai, Real-time magnetic length-based vehicle classification: Case study for inductive loops and wireless magnetometer sensors in Oklahoma state, *Transportation Research Record*, 2672(19) (2018): 102e111. https://doi.org/10.1177/0361198118791612.
43. F. Kurauchi and J.D. Schmöcker (Eds.), *Public Transport Planning with Smart Card Data*; CRC Press Taylor and Francis Group: Boka Raton, FL, USA, 2017, p. 281.
44. K. Wagh, Ashish Chauhan, Shraddha Bhandari, Aniruddha Langhe, and Nikhil Pawar, Dynamic traffic light optimization and control system. *International Journal of Future Generation Communication and Networking*, 13(1) (2020): 472–481.
45. B. Sumathy, L. Sundari, S. Priyadharshini, and G. Jayavarshini, Vehicle accident emergency alert system. *IOP Conference Series: Materials Science and Engineering*, 1012 (2021): 012042. https://doi.org/10.1088/1757-899X/1012/1/012042.
46. Aymen Aloui, Nadia Hamani, Ridha Derrouiche, and Laurent Delahoche, Systematic literature review on collaborative sustainable transportation: overview, analysis and perspectives, *Transportation Research Interdisciplinary Perspectives*, 9 (2021). https://doi.org/10.1016/j.trip.2020.100291.
47. A.A. Nasution, F.N. Nasution, and Risanty, *IOP Conference Series: Earth and Environmental Science*, 562 (2020): 012012. https://doi.org/10.1088/1755-1315/562/1/012012.

6

IoT and Big Data Analytics-Based Intelligent Decision-Making Systems

N. Sudhakar Yadav
VNR Vignana Jyothi Institute of Engineering and Technology, Hyderabad, India

Sreenivasulu Gogula
Research and Development at ACE Engineering College, Hyderabad, India

Ganti Krishna Sharma
ACE Engineering College, Hyderabad, India

Ch. Mallikarjuna Rao
Gokaraju Rangaraju Institute of Engineering and Technology, Hyderabad, India

D.V. Lalita Parameswari
G. Narayanamma Institute of Technology and Science (For Women), Hyderabad, India

CONTENTS

6.1 Introduction 101
6.2 Literature Review..104
    6.2.1 Necessity of Integrating IoT and BDA in Intelligent Decision Making 110
6.3 Proposed Work..111
    6.3.1 Convergence of Big Data and IoT in Smart Urban Transportation System 111
        6.3.1.1 Supervised Learning Algorithm 112
        6.3.1.2 Unsupervised Learning..113
        6.3.1.3 Semi-Supervised Learning..113
        6.3.1.4 Reinforcement Learning..114
    6.3.2 Decision-Making System Using Ensemble Learning (DMEL).....................114
6.4 Results and Discussions...115
6.5 Conclusion 116
References...117

6.1 Introduction

Whenever an information can be recorded, the term "data" is used; nevertheless, when the fact is used in a certain context, the term "information" is used. Information that is transmissible and storable, as well as information that can be acquired and analysed, is

DOI: 10.1201/9781003217404-6

known as data in the computer industry. The most prevalent type of data representation is tabular, which is made up of columns and rows. Large amounts of data were available in a variety of ways even before the computer was invented, including accounting, trade, calendars, astronomy, and taxation, to mention a few examples of applications. As soon as the data is gathered and processed, then organised and examined, it is changed into information. The knower's understanding and interpretation of knowledge is dependent on his or her own interpretation and understanding. During the previous three decades, data are swelling at a wider rate and heterogeneity, as well as speed and variety of data creation, which has been matched by an increase in data generation speed and variation. Data collection engines have been fuelled by information technology (IT) advances and developing technologies, which have amassed vast quantities of data and turned it into a valuable commodity over the course of the last several decades. A new age of IT has begun with the introduction of the Internet and the subsequent proliferation of new data sources. Data management has become a fundamental need for archiving and evaluating individual and organizational data in the digital era as just a result of widespread use of IT and the introduction of the Internet in practically every aspect of life. The basic blocks of data analytics are the transformation of data to knowledge, followed by the renovation of that data into understanding. To reap the full and actual advantages of raw data, it is necessary to have a data analytics system in place. It is becoming more popular. Many of the prediction systems in traffic management are based on estimated arrival time [1].

Machine learning algorithms are comprised of a set of algorithms that recognise patterns in data and predict future occurrences in order to assist decision-makers in their decision-making processes. Supervised and unsupervised are the two categories of machine learning algorithms. We use this strategy to train the computer by labelling data with the desired outcome and then using that labelled information in the computer's learning process. Tagged data is used to learn, and this learning is then used in order to predict the outcome of unexpected data, as seen in the following example: Other applications of supervised machine learning include classification and regression, among other things. Both the input and the output are supplied in this kind of learning. The computation required to carry out this type of learning is quite cheap. Because there is no need for labelling, unsupervised learning necessitates the absence of any supervision. In order to keep up with the increasing amount of data being created at a quicker rate, data processing and integration get more difficult, which adds to the total complexity. All of these factors are driving forces behind new technology development and the creation of data management systems, as well as the development of new technologies.

Because of its ubiquitous availability and ability to transmit data in the shortest and feasible period of time, the Internet has risen to be among the greatest important influences in markets in recent years, and it has become an essential part of it. Thanks to the efforts of many individuals, organisations, and governments, it has expanded from an advanced research networks with again several terminals in the 1960s to a widespread international network with over a billion active users today [2]. IoT is the furthermost current technological advancement in the Internet sector, and it survives for enormous number of applications. In the year 1999, the great person named Kevin Ashton coined the word "Internet of Things" to describe the medium that connects physical things to the global computer network. "An interconnectedness of specialised tangible items (things) having integrated technology for detecting or communicating with their intrinsic status or external influences," as per Wikipedia [3]. Paper [4] describes this by means of "a network of interconnected intelligent material objects (sensors, gadgets, machinery, commodities, and items) as well as the internet and applications". Through the introduction of Internet of Things, it stands now possible to associate the physical world to the eco-system by means of the Internet. Equipment that were originally designed to perform certain real-world roles in an idyllic world are nowadays aggressively participating in the framework in a collaborative way, which is a step forward. All physical items can link to everything and everywhere, perhaps via a network or media and can deliver any kind of service [5]. Paper [6], "The IoT is an assortment of networked manoeuvres," adding "the Internet of Things is an evolution of the underlying conceptions of pervasive computing, intelligent systems, and wearable devices." "A networking of inter-connected technology with computational competence, automated and chromatic devices, or individuals designated with tags and the potential to transfer facts via a system in the exclusion of interaction between humans or machines," according to the definition, explain [7] in their definition of IoT. It is, at its core, a critical evaluation of the Internet through which learning between objects may be achieved [8]. According to the foremost management literature, the necessity to understand how this technology impacts the notion of Business Process Management (BPM) at its most basic level is becoming more crucial [9]. Everyone agrees that there is a rising need for new ideas in the Internet of Things to be properly examined [10]. Through the swift rise of IoT and its applications in the business world, this emerging field of study is swiftly becoming a distinctive and exciting one. As per the recent estimates, this adoption of smart items will surpass 50 billion by 2020. The Internet devours been strapped into the receptacles of the typical individual along through utilization of communication and interpersonal media. As a consequence, more and more individuals are talking about data, particularly data analytics, data security, and data science, among other topics. The outcome is the availability of low-cost commodity technology for the purpose of creating an expandable data processing platform that is capable of managing the volume, diversity, and rate of data that is linked to the platform. These are only a handful of the well-known businesses that have made major contributions to this field. Others include Microsoft, SAP, IBM, Cloudera, HP, and Oracle. It is predicted that by 2020, data quantities would have reached up to 40,000 exabytes, International Data Corporation (IDC) reveals. According to the researchers, if the data is correctly analysed using Big Data technology, a further two-thirds of it may be deemed usable for decision-making purposes. The researchers also raised attention to a prior issue in which a substantial amount of vital information had been omitted during a research project.

Making decisions on how to do that is a basic biological trait that underlies all human interactions with their surroundings. The fact that humans can make both excellent and bad decisions is apparent, and scholars are debating the best effective methods for guiding (or assisting) people to a "good" decision. There are three categories of possibilities: structured, semi-structured, and unstructured. The structured choices are the most common. This categorization may assist us in determining the most effective ways to assist individuals in making choices. The use of decision help is not required in structured choice problems since the optimal answer is well defined and may be reached without it. Consider the situation in which you must choose the shortest route between two points in time. Using analytical methods, it is possible to provide a precise solution to this issue. Unstructured decisions, such as selecting a spouse, may be considered while making a career or financial decision. The semi-structured issue category encompasses a broad variety of situations that although they normally have some agreed-upon characteristics, need human input or preferences in order to make a conclusion based on a certain set of criteria. Issues that have those consented characteristics but require human control or preferences in order to make an informed decision based on a set of criteria fall under this category of difficulties. Researchers advise us all that perhaps a full understanding of decision-making is critical to making optimal use of it and reap from artificial intelligence [11]. Artificial intelligence may mimic human decision-making to some extent, and recent AI breakthroughs have shown excellent potential in assisting and supplementing rational decision-making, predominantly in current and intricate scenarios. Ensemble learning approaches are becoming progressively crucial in making intelligent decisions as well as behaving more intelligently. A framework is proposed to address the challenges with the aid of random forest and extreme machine learning processes. The recommended system is designed on converged Big Data-IoT to make intelligent decisions with no need for manual assistance.

6.2 Literature Review

As part of this section, we will look at some of the current literature, which will highlight the many ways in which organisations are employing big data for analysis and decision-making. After describing the objectives of our investigation, we uncovered the keywords "Big Data" and "Big Data and Decision Making." Other keywords, such as "Big Data analytics," were also discovered. In addition to reviewing research articles published in reputable journals, conference papers, and web-based sources, we also chose the most relevant studies for further examination. High-quality research papers have been selected for inclusion in this collection via the use of the databases Scopus, Science Direct, and Google Scholar. Several articles that are connected to the issue have been picked based on the keywords that were found and added into the database.

First and foremost, urban areas in India, as well as urban areas around the world, are changing at an alarming rate [12]. Because the infrastructure has been planned in advance, it is not a decision made on the spur of the moment. When the Smart City concept is implemented in a planned city, the consequence will be that every activity taken in the city will be watched and controlled by technology, according to the notion provided. When it comes to IT, IoT is a relatively new expertise that partakes the potential to be used in order to realize the goal of constructing a smart city. Building single-handedly will not be adequate; they will also need to retain and maintain their unique character. Another task that must be digested and completed is the task of being honest and sincere in one's actions. When it comes to developing a smart city in India, there are several challenges to overcome, both implicit and explicit, and these obstacles should essentially be addressed in demand for the work to be successful. An efficient smart city model cannot be implemented due to the fact that each municipality is distinct in its own way. The building of a prototype, on the other hand, is necessary in order to generate a logical design for Smart Metropolises utilising the Internet of Things. The second and third contestants [13], using massive amounts of data gathered from highways, the authors want to investigate predictive technology-based road safety as a means of improving traffic safety. It is the purpose of this study to investigate aspects of traffic management systems and technology for road safety assessments that have been successfully applied in Korea and other countries. On the subject of traffic management systems, the kinds of information that may be acquired and the ease with which it can be retrieved are studied. After considering the results, it is necessary to consider the limitations of present technology and management practices. Several relevant techniques and traffic management systems have been researched utilising fundamental physics depending on the distance, velocity, and other variables, as well as past event information; nevertheless, they have failed to take into account a range of unique aspects and statistical information. As a consequence, technology for maintaining roads and traffic must be developed that uses a huge set of data sources in actual environments, such as traffic data, meteorological data, and pavement condition data, as well as data analysis. Using data from many types of traffic management systems, it will be feasible to make more reliable methodologies and systems for traffic safety regulation. The third group includes [14], in order to get things started; this article will summarise the present status of the issue and make predictions about the key functions that will be performed in the future, among other things. This chapter emphasizes the use of big data analytics in smart urban then examines how urban areas may contribute to the transformation of our way of life. Lastly, it discusses some of these developing technologies' potential downsides, such as their ability to deceive and compromise our personal information, security and privacy.

However, while big data has the potential to be extremely beneficial, this even poses a serious threat to our individual privacy and security. Business organisations, governments, and individuals may acquire and sell data, which is more than simply another piece of information. Clients are encouraged to think on how the huge volumes of data acquired about them could be used to restrict their freedom, and that it's up to the discretion how much information they are comfortable sharing in this piece. Briefly stated, when a democratic society upholds and protects the rights and freedoms of its members, the benefits much outweigh the risks. This technology has a wide range of applications, including power management, energy consumption management, and distributed storage coordination, to name a few. Advanced sensor infrastructure is required to be widely used in the future of intelligent grids due to its cost effectiveness and ease of implementation. Specialists can use the in-depth knowledge gathered from the acquisition and analysis of such data to optimize power grid functioning and so obtain superior performances. Based on an extensive study, this article includes practical proposals that could be implemented in the forthcoming of the smart grid as well as the Internet of Things. The tactics for dealing with enormous quantities of data collected by sensors and metering for the purpose of data processing are also covered.

Hadoop is one of the software technologies that have been successfully used in this way in order to store and analyse massive data volumes on consumer-grade hardware. There are many different forms of amalgamation of R with Hadoop for processing huge datasets, and the author addresses them all in this post. These include integrations with Streaming, Rhipe and RHadoop, just to mention three examples.

Paper [15] attempts are being made by the author to identify what exactly defines a smart city. Entities that have been created to characterise this vast field of study have defined these cross-sectional studies as they see fit. As a result of the interaction of these entities with continuously changing social, technological, economic and political variables, possibilities for the ongoing refining of the idea of smart city are created every day. While most emergent qualities are contextual in nature, they have an influence on both the sorts of data that may be collected in cities and the ability of cities to develop sophisticated information systems. In the field of computer analysis, large-scale data integration is a well-known problem that includes integrating massive amounts of data that have been generated. The diversity and complexity of data streams in a smart urban implies that a scheme might remain particularly well suited for intensifying a wide range of drivers and dynamics, as well as offering adaptable reaction methods to these drivers and dynamics. However, due to the extremely unpredictable and chaotic nature of these systems, it is only natural for them to need stabilisation in order to work correctly. Many other types of criteria, such as parametric, entropy, and anthropic concerns, have been proposed in various research projects. Because of the large variety of materials and variables that influence the system's drivers, it has been proposed that attractors developed from complex systems be utilised to characterise smart city settings via the various linked big data and information systems. Furthermore, as the trading volume of traffic increases, typical software approaches for storing and analysing massive datasets are becoming more complex [16, 48]. It's critical to have a set of storage devices capable of concurrently data storage and processing tremendous amounts of data in order to store and analyse such large amounts of data. Among the frameworks available are Hadoop, which provides a dependable cluster storage facility, and the MapReduce framework, which provides functions for efficient parallel processing. Hadoop stores massive quantities of data in a distributed way, utilising special file format known as the system distributed Hadoop records. Processed transmitted data can be acquired fast and easily using MapReduce, allowing end users to perform traffic analysis or health data analysis to obtain crucial predictions about their surroundings and similarly kafka and storm used to process the big data [17].

According to the author for Paper [18], businesses located in smart urban are having problems establishing themselves and so falling short of their estimated potential. This paper considers how this is the fact, and it proposes a method for smart cities to emerge as a result of the exploitation of large data through the idea stores of application programming interfaces. In order to do this, it is important to first study the many actors involved as well as the overall ecosystem. Next, a plausible route to commercial scale is proposed within the context of that setting. An examination of the ICT technologies now accessible is also included, and an application of intelligent sustainable city is utilised to show all of the findings. Throughout this article, the writers make two important discoveries that will aid in the long-term development of the intelligent city and its residents. It is essential to first and foremost create separate smart city departments (or equivalents), much as IT departments are generally recognised today. This will allow for a clear distinction to be made between the political component of city development and the underlying technology. Second, a three-phase implementation approach for smart cities is required to be successful. A huge quantity of data is generated by cities on land use, the environment, socio-economics, energy, and transportation, among other things. Incorporating an integrated management perspective and conducting in-depth data analysis can provide answers to a wide range of scientific and policy questions, as well as planning, governance, and business issues. It can also provide decision support in an environment that allows for more informed decision-making in a more efficient manner. With the use of a cloud-based analysis service, this paper offers a conceptual and applied viewpoint on huge data management and data analytics for smart cities, as well as a service proposal. The prototype was built employing Hadoop and Spark, as well as the comparison is made to each other in order to compare the findings. While analysing data from the Bristol open data collection, the service discovered a correlation between chosen aspects of the urban environment. Experiments with Hadoop and Spark were carried out, and the results of these experiments are described in this paper. Urban data, described by the author as "data for cities that is always labelled with place and period," is defined as "data for cities that is always labelled with place and time." The author describes large amounts of data as "data for conurbations that is always labelled by means of place and time." He emphasises that sensors transmit the vast majority of the data, and that this signifies a significant shift in the forms of content we have on what occurs what and how in communities. He feels the focus is shifting away from long-term strategic planning and toward short-term considerations of how cities run and it can be regulated as a result of the massive amount of data being created. Thus big data will become an abrupt technology with a cradle of knowledge for people altogether around the world in the future, regardless of their location or time zone. At the end of this study, an observation to six months of record commute smartcard individual's trip in the London public transit systems is presented to highlight the need for new theory and analysis. We shall discuss the strategy of intelligent transportation structures in the light of the rise of smart cities in this article. It provides an overview of the smart transport system depending on the level of vast volumes of data and relies on traffic construction programmes' wisdom to provide insights into the process. The city of Lanzhou, China, was evaluated and developed using this approach. In order to adequately investigate and analyse the vast quantity of GPS data that was collected, researchers developed a mathematical model. As a consequence, to anticipate traffic flow, it is necessary to apply a clear graph analysis of city traffic flow. Finally, intelligent transit is recognised among the characteristics of a smart city, which is exactly what we desire. The term "smart cities" has expanded prominence in academe, corporate, and public towards portray metropolises that are progressively made up of this and carefully monitored by distributed network computing, while also having a government and economy propelled forward by the ideas of intelligent people who are constantly innovating, creating, and entrepreneurship. First and foremost, it is the first of these that is the subject of this chapter, this demonstrates in what way metropolises are increasingly fortified with digital gadgets and facilities that generate "big data" through a series of instances. According to some proponents, the intelligent city is a type of data that enables critical analysis of urban life, the development of innovative urban management models, and the generation of new ideas for efficient and sustainable cities that are competitive, productive, and transparent, all of which can be imagined and implemented. It is possible to build smart cities via the integration of information and communications technology, which allows for more effective control of urban infrastructure and the management of urban infrastructure to be achieved. Wireless networks, the Internet of Things, digitization, large amounts of data engendered by biomedical Internet sites, social media, Navigation system, and geographic information systems (GIS), among several other applications, are all significant source of data for the development of a smart city with effective management governance.

Identifying and classifying groups of unstructured and structured data, integrating them together into cloud environment utilizing adequate hardware and the software, and developing user-friendly interfaces will be a profitable method. The long-term administration of smart urban will be aided by retrieving the right information system and user. The proposed approach, while using the Hadoop architecture, preserves network data security over a large mobile network while simultaneously processing massive volumes of data commuters. There are several experimental programmes now underway to make cities smarter. The research also examines the benefits to human wellbeing that will arise from these activities, which are currently unknown. Additionally, the research investigates the different obstacles that may arise when seeking to apply Big Data solutions to the enlargement of smart metropolises. The continual increase in the amount of information produced has the potential to benefit both government and commercial businesses. Within the huge quantity of data, there are a variety of points of view to consider, as well as some really essential facts. A detailed understanding of the challenges that develop as a consequence of the generation of large amounts of data, on the other hand, is required. In this day and age, the whole world is linked.

In this chapter, we will cover a variety of aspects of intelligent cities, as well as the obstacles and concerns that these developments bring with themselves. Furthermore, the significance of big data analysis, as well as the tools and technologies that are now accessible, are discussed. The concept of an intelligent city was born out of an effort to increase the overall standard of living for those who live there. The key concept is the approach to information network services for every location. These areas include health care, education, transportation, and power distribution, among other things. There are several problems and requirements as a result of these expectations. For the purposes of this research, the most important Information and Communication Technology (ICT) issues related with adoption of smart cities will be identified. As a consequence of the study, a high-level architecture for intelligent cities has been presented, which recognises the need of effective data collection, storage, and retrieval, as well as the efficient provision of network resources, in order to achieve smart cities. Furthermore, the proposed framework outlines how different stakeholders connect with one another and deliver services to individuals, and it is built on a hierarchical model of data storage to do this. In terms of reaching the requisite degree of sustainability and improve living circumstances for all residents, most governments were exploring the notion of smart metropolitan development.

By way of an effect of a rising population and the need of having an automobile as a basic prerequisite for living, traffic congestion is developing at an uncontrollable rate. The deployment of efficient and effective traffic management technologies is essential if we are to prevent traffic chaos in the foreseeable future. For aid in the reduction of traffic management difficulties, particularly in India, the programme provides users with a snapshot of traffic conditions in different regions of the country, allowing it to move more quickly and avoid traffic jams. We will investigate the major data analysis tools that can be used to analyse the massive amounts of data accumulated from transportation systems and expected outcomes that could be used to establish a comprehensive traffic experiment and much more effective organizational policy that can significantly reduce vehicular traffic delays and mishaps completely while also maximizing the efficiency. Identifying the causes that are linked with a vehicle accident and the factors that determine traffic is among the most important tasks in the evaluation of crash data. The authors of this research employed K-modes clustered as a prior work while segregating 11,575 vehicle mishaps on the roadways of Dehradun in India during the year 2009, and the year 2014. Based on the empirical results, they established a model for segregating traffic fatalities on Indian highways. Lastly, depending on all of the statistics completed thus far, they compare the performance of cluster analysis. It is conceivable that difficulties may occur as a result of the massive volume of information that is being processed. As a result, the author suggests that the Hadoop design be employed to address this problem. The MapReduce framework, which is built on the parallel processing architectural paradigm at its core, may be used to make data available in a number of places by distributing it over a distributed network. To accomplish so, a system must be devised that records the journey time of cars as they pass over the reader's roadways, determines an overall transit time, and then distributes that data to various toll gathering locations across the globe. As a result, the author proposed that data frames be processed in a distributed manner across several locations, using the Hadoop architecture. Through the use of MapReduce, the author was able to do this.

The Internet of Things has evolved in many facets of life since the advent of artificial intelligence, and it has piqued the interest of academics who are attempting to create a new paradigm of living norm. This invention has been widely embraced across the world as a means of making life simpler as a result of the rapid proliferation and widespread usage of different smart digital devices. The use of AI-enabled gadgets makes them further intellectual and proficient of performing a particular job, which hoards a significant amount of time and money. Because of their low cost and adaptability, the Internet of Things, mobile, and network applications offer a superior solution [19]. The Internet of Things' primary role is to connect users to readily available resources while also providing them with dependability, effectiveness, and smart service. While it derives to elegance, the Internet of Things puts sensors with seamless operations, a remote server, and a network together in one package. The plan is extensive in terms of provided that perceiving with multidimensional formations and delivering core therapy suggestions [20], among other objects. There are many central focal uses for the Internet of Things (IoT) that enhance the quality of life.

The function of decision-making in AI-enabled and Internet-of-Things systems is significant in and of itself. In paper [21], proposed the notion of a business model for firms that are using IIoT technology. The strategy has been designed in order to assist conventional firms in their transition to the digital market environment. Paper [22] provided a wide-ranging overview of the anticipation, development, and extenuation of ransomware prevalent the Internet of Things scenario. Paper [23, 47] has proposed a technique for studying the fruition of the Internet of Things, expertise related with cloud and smart urban, and then concentrating on the expertise of the Internet of Things and cloud. Paper [24] proposed an approach for accessing cloud-based services for IoT devices.

Big Data may then be used to analyse and extract the necessary data in order to generate the information that is needed. While Big Data may be used to analyse enormous quantities of data in real time, it can also be used to store the information using a variety of storage methods. This makes Big Data very useful when it comes to using the capabilities of the IoT and extracting information. Another point of differentiation is the method of data gathering. For example, Big Data collects information on human behaviour in order to make forecasts or uncover patterns of behaviour. The data provided by the Internet of Things, alternatively, is machine-generated and is used to achieve optimum performance in equipment or to decide predictive maintenance. They do, however, function in concert with one another. The Internet of Things collects real-time analytics data to assist with real-time decision-making. A big data solution, which serves as a storage solution for predictive analytics to foresee future issues and develop solutions, assists in this role. When IoT and Big Data work together, they may analyse inputs to reveal hidden connections, previously undiscovered patterns, and uncover new trends in your data collection. Because of the interdependent relationship between the IoT and Big Data, as IoT continues to expand at a fast pace, there will be more strain on conventional data storage, resulting in the development of more inventive Big Data solutions. As a result, organisations will be obliged to modernise their technologies and systems in order to keep up with the increased demand. At the end of the day, Big Data and IoT have identical objectives and depend on one another to accomplish them by transforming data into information that can be used by enterprises. Using real-time IoT insights mixed with long-term Big Data analytics, for example, may be utilised to provide a more comprehensive view of a company's overall performance over time. Specific sectors will enjoy the advantages of these technologies as they continue to grow and advance. Sectors such as the automotive, shipping, and haulage industries will gain from the use of IoT and Big Data. As a result of the insights and information gained via technological advancements, the analytics process will be made more efficient and easy, resulting in cost savings as well as the promotion of informed decision-making and forecasting.

The survey report sought to provide academics with relevant information about the characters that data-driven methodologies can play in smart metropolises and transit [25]. This paper's goal was to familiarize scholars about emerging advancements and to provide relevant perceptions into how techniques can be used in place of SC planning and management. An ideal location technique for an intelligent traffic logistics facility is proposed in the paper [26] with the help of block chain and IoT technologies [26]. To empower IoT systems to offer input, this approach integrates a number of sophisticated processes.

The concept takes the advantages of a blockchain service's irreducibility to encourage the creation and deployment of blockchain logistics solutions. This technique is used to monitor the development of goods transportation across the distribution network. The objectives of the paper is to summarize the key tendencies and open problems of implementing Intelligent systems for the establishment of effective and resilient smart cities by reviewing investigation literature [27, 28]. The paper begins with a review of the core tools presented in the writings for the deployment of IoT structures, followed by an analysis of the primary smart urban strategies plus methodologies. The paper discusses large data handling and analytics in intelligent smart settings, with a focus on smart logistics and transport systems, and recommends research paths that could be a turning point in the next years [29].

Abundant researchers had showed an interest in Machine Learning methodologies, and their approaches had aided in the resolution of related problems. The paper gives audiences an insight of IoT technology and the use of several authors' machine learning methods [30]. The goal would be for future scholars of being capable of recognizing outstanding difficulties and problems and use various methods to address them. This paper introduces on emergence of new apps that use data from IoT devices to provide meaningful results and making things smarter using machine learning techniques. IoT solutions are helping to propel the worldwide IoT in Intelligent Transportation System forward. The deployment of IoT in ITS with an integration of sensors and analytics has resulted in effective handling of the current traffic issues. IoT-based ITS assists in the automation of smart traffic supervision in cities in all aspects [31]. An agent-based traffic management system deploying effective light control and deviation strategy is proposed and examined [32]. Regression techniques are used to forecast the traffic flow in a short span and thereby this also handles the scenario in holidays and weekends when the traffic is quite high [33]. Extreme machine learning approach will assist the machine to take optimal choices; thereby random forest shall suggest the better decisions. With the support of these mechanisms, an automated intelligent decision-making system shall be established for better urban services. This chapter lays forth a roadmap for dragging these mechanisms in place in attempt to develop automated intelligent decision-making systems that are also optimal at all times.

6.2.1 Necessity of Integrating IoT and BDA in Intelligent Decision Making

In order to enhance urban service delivery while also incorporating the benefits of Internet-of-Things-based smart cities into their work, industry and academic specialists have devised a number of solution methodologies. As a consequence, a large amount of experimental research and testbed-based studies were carried out in order to uncover the obstacles associated with smart cities and to seek the solutions to the concerns that were discovered. Despite the fact that data obtained from deployed sensors is crucial in real-world smart city deployment, data acquired from deployed sensors was not evaluated in this research, despite the fact that they are critical. The computational architecture includes radio frequency identification (RFID) sensing in order to increase the traceability of cities and to enhance the traceability of their residents. In a heterogeneous environment, this architecture collects information about the user from a variety of sources in a reliable and transparent manner. The authors of this article included people's mobility patterns into their work in order to minimise deviations from actual movements during transportation route design, as described in the paper. Furthermore, because of the many benefits that

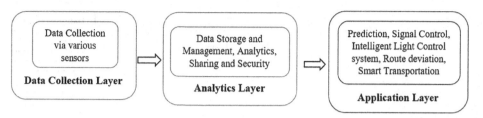

FIGURE 6.1
Role of data analytics in ITS.

cloud computing offers, a considerable number of literature studies devour cloud-based development. Cloud-based service assists in an improved smart city [34].

SmartCityWare is a virtual environment that enables developers to design smart city applications in a virtual environment. It is envisioned that every function in this cloud-based employment is a service given by a smart urban. Several types of clouds and gadgets, as well as fogs, are used to distribute the services in this manner. BDA tools are reliant on the essential characteristics of BD, which are referred to as VVV characteristics (volume, variety, and velocity). The role of data analytics is presented in Figure 6.1.

Urban inhabitants' quality of life may be improved via the use of autonomous intelligent decisions that are derived from data generated inside the city's network. However, real-time Big Data processing is very advantageous in the design and administration of cities, and it is particularly beneficial in the transition of these cities into smart cities. The BDA, on the other hand, ensures the dependability of penetrating and essential decisions, which, alternatively, increases the overall effectiveness and profits of local government bodies. This chapter employs extreme learning machine algorithm and random forest for improved decision-making process.

The following sections break down the chapter into its several sections. In Section 1, we discuss the introduction of big data and Internet of Things, and in Section 2, we discuss the corresponding work that is covered in the first section of this document. It is discussed in Section 3 how hybrid intelligent decision-making systems in smart cities may be implemented by combining the Internet of Things with Big Data analytics. These are the elements that make up the system: Decision-Making System Using Ensemble learning (DMEL). Results and discussions are included in Section 4 of the suggested architectural design. Section 5 concludes the work.

6.3 Proposed Work

It is recommended that a smart city framework, which contains BDA, be separated into three tiers, which are categorised as follows: data aggregation, data management, and service management.

6.3.1 Convergence of Big Data and IoT in Smart Urban Transportation System

The convergence of Big Data and IoT in current traffic management is required to provide a versatile, smart, and smart framework for smart transport systems. This chapter delves

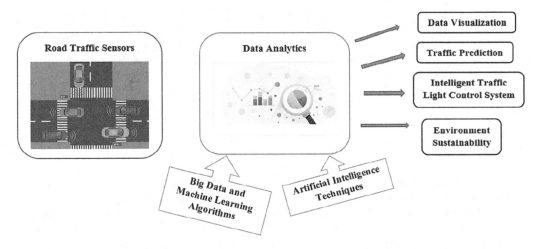

FIGURE 6.2
Framework for smart intelligent transportation system.

into the framework's strategies as well as the function of transportation schemes in a smart urban. The characteristics of IoT are crucial in acquiring and processing key data. Data analytics, on the other hand, aids in the discovery of insights and the extraction of features from collected data. As a result, they influence the development of intelligent prediction as well as decision-making systems with the potential to decide what is best possible based on the circumstance. Multiple sensors are commercially available to help with data collection. The framework for smart transportation deploying data analytics and sensors are illustrated in Figure 6.2.

Machine learning is perhaps the most prominent model and analytical framework in Big Data groups since it varieties this easier to identify sketches and insights since massive amounts of facts. In addition, the machine learning algorithms are widely used in ITS fields to undertake data analysis. Be contingent on the comprehensiveness of the records presented aimed at learning, the machine learning methods can be classed as unsupervised, supervised, and the reinforcement learning approaches. Advanced deep learning techniques devise recently stood implemented in ITS, thanks to the rapid advancement of Artificial Intelligence.

6.3.1.1 Supervised Learning Algorithm

This learning method takes the trained dataset [35]. Using input data and targeted outputs, the models train the function as well as mapping among input information with target outputs (labels). The learnt model and also the input can certainly predict the unobservable outputs. Decision trees, support vector machines, neural networks, and linear regression seem to be the most frequently applied supervised machine learning techniques in ITSs.

The goal of regression would be towards describing the relationship among dependent variable too one or much added target variable. A Linear regression is a meek, reliable, simple-to-understand, and simplified technique. Linear regression has been shown to be useful in a range of ITS contexts, along with traffic prediction [36], traffic speed estimate, and transportation route optimization [37], due to its simplicity. Classification algorithms in which decision trees are certainly a decision-making methodology which utilises a tree-like graph structure to ideal options and its likely outcomes [38]. Since it is versatile, durable, and simple, these are frequently applied in many ITS applications, such alike road accident detection, smash severity analyses, and mode selection.

A prominent form of form of incessant and strong supervised method employed for both regression and classification is an artificial neural network (ANN) [35]. When there are enough hidden and output of computation elements and testing dataset, an ANN could acquire slightly non-linear relationships among input and resulting data. It's also been employed as a data modelling framework in ITS applications like traffic flow prediction [39], trip time forecast, traffic accident monitoring, and leftover public parking prediction [40]. An approach that is quite popular in supervised learning techniques are support vector machines, which employs labelled data for classification and regression problems. Among all of the data analytics modelling strategies in ITS, these SVMs have caught academics' curiosity. It's been used in transit time estimation, prediction of vehicle arrival time, and traffic collision identification [41]. SVM has been used to predict traffic incidents in [41], and this is a prominent sample of supervised learning being utilised in the real-time urban transportation.

Consider the training subclass $\{(a_1, b_1), (a_2, b_2), ..., (a_i, b_i)\}$, where a_i represents the input variable of the model trained which encompasses several factors of road traffic flow rates namely volume, velocity, occupancy and consequently, b_i is the class label of a_i. Adding to this, a kernel function $k_f = (a, a')$ conferring to the support vector machines classifier philosophy, the equation to obtain the support vector α_i shall be resulted as

$$max_{\alpha_i} -\frac{1}{2}\sum_{i=1}^{l} b_i b_j \alpha_i \alpha_j k_f\left(a_i, a_j\right) + \sum_{i=1}^{l} \alpha_i \text{ subject to } \sum_{i=1}^{l} b_i \alpha_i = 0 \quad (6.1)$$

The decision function $df(a)$ to calculate the label for the model can be attained as

$$df(a) = \sum_{i=1}^{l} b_i \alpha_i * k_f\left(a_i, a\right) + c \quad (6.2)$$

Uncertainty a is an occurrence sample, $df(a) = 1$. Else, we take $df(a) = -1$

6.3.1.2 Unsupervised Learning

To learn with no oversights is the goal of unsupervised learning, also known as clustering, is to find natural groups in untrained multi - dimensional statistics [57]. K-means would be the greatest extensively used unsupervised method for learning, with applications in vehicular traffic management and trip time estimation [42]. Paper [42] discusses a well-known example utilising unsupervised learning to forecast travel duration based on past data.

6.3.1.3 Semi-Supervised Learning

Learning in a semi-supervised environment: This category combines the two strategies discussed above. There is a significant amount of input data in this form of learning, although some of it is merely labelled. In this sort of learning, a collection of labelled and unlabelled data is used.

6.3.1.4 Reinforcement Learning

Reinforcement learning is a type of learning wherein an agent knows by trial - and - error through an interactive experience while also getting feedback itself from events and experiences. Intelligent robots and intelligent agents actively optimize its effectiveness by evaluating the idyllic behaviour inside a given framework. This type of problem getting handled defines reinforcement learning, as well as the strategies are classified under reinforcement learning algorithms. Throughout maximising the aggregate total reward, the agents determine the best actions to perform depending on the latest condition of the challenge environment. The practice of recurring this phase of the problematic is notorious by means of a Markov Decision Process.

6.3.2 Decision-Making System Using Ensemble Learning (DMEL)

The extreme learning machine is a novel and fast SLFN learning method wherein the inputs weights and hidden units biases were randomly chosen [43]. The heuristic issue posed by learning the properties of an extreme learning machine system can be solved analytically, producing a closed-form approach requiring just matrix multiplication as well as inversion. Like a way, unlike the standard SVM formulation, the process of learning can be accomplished rapidly without the usage of an iterative technique such a backpropagation-neural network (BP-NN) or the approach to a quadratic programming challenge. In extreme learning machine, the hidden influences and inputs weight (which connect the input layer to the hidden layer) are determined at random, whilst the output weights (which connect the hidden layer to the output layer) are calculated using Moore-Penrose MP generalised inverse [44].

Decision-making system using ensemble learning is a decision-making method that uses a group of models, including one that was developed using machine learning approach on relevant data. To deliver the best results prediction, DMEL integrates various models in a distinctive manner. They've being widely used in machine learning and pattern recognition since they've been found to surpass solitary models and considerably improve accuracy rate. DMEL is used for statistical data, large data quantities, insufficient information, and data acquisition, among many other things. Further information is available at [45]. Methodologies neural network-based, such as Boosting , Bagging, and AdaBoost, that necessitate learning process tweaks and utilise an amalgamation of constraints, including certain algebraic combination; voting-based approach; decision templates; and techniques based on neural networks, such as Boosting, Bagging, and AdaBoost. The method used in this study is arithmetic averaging, which is simple and allows for impressive results. The proposed strategy is illustrated in Figure 6.3.

Random forests, often called random choice forests, are an ensemble learning approach, regression, as well as other processes that operate by training a vast number of decision trees and afterwards outputting the class that is the mean prediction (regression) of the individual trees. The pseudocode of this method is illustrated below.

Step 1: Pick 'r' distinct attribute from the total 'm' attributes.

Step 2: Check r<m.

Step 3: Calculate 'n' node by applying the finest split strategy.

Step 4: Distribute the node into offspring with the finest split.

Step 5: Repeat the steps from 1 to 3 until nodes are reached to be 1.

Step 6: Build forest by reiterating the steps from 1 to 4 for 'i' times resulting in 'i' trees.

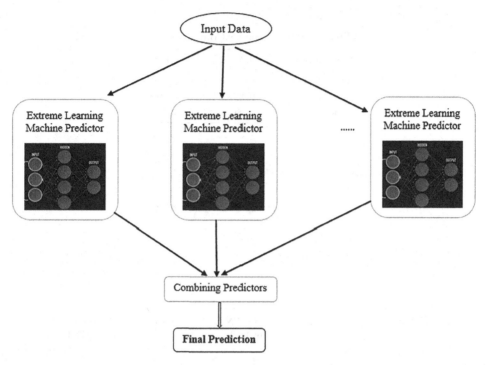

FIGURE 6.3
Prediction using the proposed approach.

6.4 Results and Discussions

The framework we constructed can be evaluated using a variety of factors; it provides us with a clear picture of its effectiveness and how it predicts the outcomes. There are numerous metrics for regression models to predict quantitative value, such as RMSE, MAE, and MAPE [46]. The three preceding Metrics have been used to assess accuracy of the model in this paper. The AIC was also used to compare models in besides these indexes. The following are the formulae for these performance indicators:

Root Mean Square Error (RMSE) can be computed with:

$$RMSE = \sqrt{\frac{1}{N}\sum_{i=1}^{N}(a_i - \hat{a}_i)^2} \tag{6.3}$$

Mean Absolute Error (MAE) can be obtained as:

$$MAE = \frac{1}{N}\sum_{i=1}^{N}|a_i - \hat{a}_i| \tag{6.4}$$

where 'N' represents the number of observations in the test dataset, a_i is the reel output of ith observation and $\hat{a}_i$ is the predicted value for i^{th} observation.

TABLE 6.1

Prediction Performance Comparison

Mechanisms	RMSE	MAE
ELM	0.028779	0.019666
DMEL	0.027479	0.018949
SVR	0.189162	0.152838

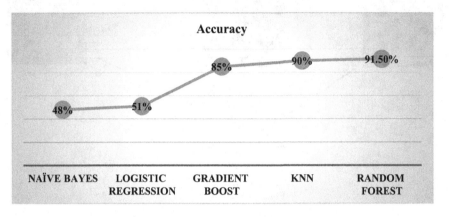

FIGURE 6.4
Performance of the models.

Ensembles approaches have been suggested in order to lessen the instabilities created by the extreme learning model. The goal is to improve the accuracy of the model's final determination. The suggested model integrates extreme learning excellent performance alongside ensemble methods' benefits. We first adapt the extreme learning model to our training data N times, where N represents the number of ensembles. The number of neurons in extreme learning hidden layer, and also N, are affected by the degree of the results. Following that, the ensemble's outputs are algebraically integrated by averaging, which reduces the danger of selecting a poor prediction. Table 6.1 presents the comparison of the performance of prediction techniques.

An Akaike Information Criterion (AIC) score evaluates the model's complexity and effectiveness; the DMEL model seems to have the least training costs, as per AIC values are computed. This demonstrates that the technique is superior to SVR, ARIMA, and ANN models in terms of quality. The hybrid model, as shown in Figure 6.4, seeks to reduce the fluctuations created by the extreme learning model, which improves the quality of the model performance.

6.5 Conclusion

Smart cities have enhanced the quality of life of urban communities via sophisticated advances in healthcare, transportation, utility management, and many other areas. A smart city is a consolidation of different smart components, and therefore has improved the eminence of lifecycle of metropolitan communities. It is essential for the employment of an intelligent smart urban that all basic smart components work seamlessly together and are integrated in a cohesive manner. The chapter discusses the importance of intelligent decision-making systems in driving the smart city further towards automation and modernization. The method was calibrated using a methodology based on extreme machine learning and random forest-based ensemble approaches. Furthermore, the importance of Big Data and IoT convergence in resolving decision-making obstacles were discussed. With today's improved technologies, smart city applications are likely to evolve to a new level that will mould the smart city's future. We shall work to expand the proposed methodology with an AI-based interactive approach in the future. In addition, we would like to use AI-based decision-making systems in a range of diverse applications across all sectors.

References

1. Zafar, N., & Ul Haq, I. (2020). Traffic congestion prediction based on Estimated Time of Arrival. *PloS One*, *15*(12), e0238200.
2. Kopetz, H. (2011). Real-time communication. In *Real-time systems* (pp. 167–189). Springer, Boston, MA.
3. Steenstrup, K., & Kutnick, D. (2015). The Internet of Things Revolution: Impact on Operational Technology Ecosystems. Gartner.
4. Reddy, B. Eswara et al. (2018). A centralized health monitoring system using health related sensors integrated to the mobile. *International Journal of Vehicular Telematics and Infotainment Systems (IJVTIS)*, *2*(1), 68–79.
5. Sundmaeker, H., Guillemin, P., Friess, P., & Woelfflé, S. (2010). Vision and challenges for realising the Internet of Things. *Cluster of European Research Projects on the Internet of Things, European Commision*, *3*(3), 34–36.
6. Dohr, A., Modre-Opsrian, R., Drobics, M., Hayn, D., & Schreier, G. (2010, April). The internet of things for ambient assisted living. In *2010 Seventh International Conference on Information Technology: New Generations* (pp. 804–809). IEEE.
7. Nolin, J., & Olson, N. (2016). The Internet of Things and convenience. *Internet Research*.
8. Iqbal, J., Khan, M., Talha, M., Farman, H., Jan, B., Muhammad, A., & Khattak, H. A. (2018). A generic internet of things architecture for controlling electrical energy consumption in smart homes. *Sustainable Cities and Society*, *43*, 443–450.
9. Al-Mashari, M., & Zairi, M. (2000). Revisiting BPR: A holistic review of practice and development. *Business Process Management Journal*.
10. Pomerol, J. C., & Adam, F. (2008). Understanding human decision making–A fundamental step towards effective intelligent decision support. In *Intelligent decision making: An AI-based approach* (pp. 3–40). Springer, Berlin, Heidelberg.
11. Mohamed, N., Al-Jaroodi, J., Jawhar, I., Lazarova-Molnar, S., & Mahmoud, S. (2017). SmartCityWare: A service-oriented middleware for cloud and fog enabled smart city services. *IEEE Access*, *5*, 17576–17588.
12. Ahmed, K. B., Bouhorma, M., & Ahmed, M. B. (2014). Age of big data and smart cities: privacy trade-off. *arXiv preprint arXiv:1411.0087*.
13. Ianuale, N., Schiavon, D., & Capobianco, E. (2015). Smart cities, big data, and communities: Reasoning from the viewpoint of attractors. *IEEE Access*, *4*, 41–47.
14. Bijjaragi, K., & Tijare, P. (2016). Big data approach for secure traffic data analytics using Hadoop. *International Journal on Recent and Innovation Trends in Computing and Communication*, *4*(5), 539–542.

15. Yadav, N. Sudhakar, Srinivasa, K. G., & Reddy, B. Eswara (2019). An iot-based framework for health monitoring systems: A case study approach. *International Journal of Fog Computing (IJFC)*, *2*(1), 43–60.

16. Repko, J., & DeBroux, S. (2012). Smart cities. *IMT 598 Spring 2012: Emerging Trends in Information Technology*, 1–18.

17. Vilajosana, I., Llosa, J., Martinez, B., Domingo-Prieto, M., Angles, A., & Vilajosana, X. (2013). Bootstrapping smart cities through a self-sustainable model based on big data flows. *IEEE Communications magazine*, *51*(6), 128–134.

18. Vrabie, C. (2018). Global urbanization and the need of smart cities development. *Challenging the Status Quo in Management and Economics*, 1175–1185.

19. Gierej, S. (2017). The framework of business model in the context of Industrial Internet of Things. *Procedia Engineering*, *182*, 206–212.

20. Humayun, M., Jhanjhi, N. Z., Alsayat, A., & Ponnusamy, V. (2021). Internet of things and ransomware: Evolution, mitigation and prevention. *Egyptian Informatics Journal*, *22*(1), 105–117.

21. Jiang, D. (2020). The construction of smart city information system based on the Internet of Things and cloud computing. *Computer Communications*, *150*, 158–166.

22. Ang, K. L. M., Seng, J. K. P., Ngharamike, E., & Ijemaru, G. K. (2022). Emerging technologies for smart cities' transportation: Geo-information, data analytics and machine learning approaches. *ISPRS International Journal of Geo-Information*, *11*(2), 85.

23. Yadav, Sudhakar, et al. (2021). An automated rescue and service system with route deviation using IoT and blockchain technologies. In *2021 IEEE Mysore Sub Section International Conference (MysuruCon)*. IEEE.

24. Bellini, P., Nesi, P., & Pantaleo, G. (2022). IoT-enabled smart cities: A review of concepts, frameworks and key technologies. *Applied Sciences*, *12*(3), 1607.

25. Cuzzocrea, A. (2019, December). Big data management and analytics in intelligent smart environments: State-of-the-art analysis and future research directions. In *Proceedings of the 21st International Conference on Information Integration and Web-based Applications & Services* (pp. 5–7). Munich Germany.

26. Yadav, Sudhakar, Yeruva, Sagar, Sunil Kumar, T., and Susan, Talluri. (2021). The improved effectual data processing in big data executing map reduce frame work. In *2021 IEEE Mysore Sub Section International Conference (MysuruCon)*. IEEE.

27. Dogra, A. K., & Kaur, J. (2022). Moving towards smart transportation with machine learning and Internet of Things (IoT): A review. *Journal of Smart Environments and Green Computing*, *2*(1), 3–18.

28. Yadav, N. Sudhakar, et al. (2021). Accessing cloud services using token based framework for IoT devices. *Webology*, *18*(2).

29. Muthuramalingam, S., Bharathi, A., Gayathri, N., Sathiyaraj, R., & Balamurugan, B. (2019). IoT based intelligent transportation system (IoT-ITS) for global perspective: A case study. In *Internet of Things and Big Data Analytics for Smart Generation* (pp. 279–300). Springer, Cham.

30. Sathiyaraj, R., & Bharathi, A. (2020). An efficient intelligent traffic light control and deviation system for traffic congestion avoidance using multi-agent system. *Transport*, *35*(3), 327–335.

31. Rajendran, S., & Ayyasamy, B. (2020). Short-term traffic prediction model for urban transportation using structure pattern and regression: an Indian context. *SN Applied Sciences*, *2*(7), 1–11.

32. Michalski, R. S., Carbonell, J. G., & Mitchell, T. M. (Eds.). (2013). *Machine learning: An artificial intelligence approach*. Springer Science & Business Media, Berlin, Heidelberg.

33. Sun, H., Liu, H. X., Xiao, H., He, R. R., & Ran, B. (2003). Use of local linear regression model for short-term traffic forecasting. *Transportation Research Record*, *1836*(1), 143–150.

34. Zenina, N., & Borisov, A. (2013). Regression analysis for transport trip generation evaluation. *Information Technology and Management Science*, *16*(1), 89–94.

35. Kumar, Boggula Varun, et al. (2021). Artificial intelligence based algorithms for driver distraction detection: A review. In *2021 6th International Conference on Signal Processing, Computing and Control (ISPCC)*. IEEE.

36. Vlahogianni, E. I., Karlaftis, M. G., & Golias, J. C. (2005). Optimized and meta-optimized neural networks for short-term traffic flow prediction: A genetic approach. *Transportation Research Part C: Emerging Technologies*, *13*(3), 211–234.

37. Zhu, X., Guo, J., Huang, W., & Yu, F. (2016). *Short Term Forecasting of Remaining Parking Spaces in Parking Guidance Systems* (No. 16-5060).

38. Xiao, J., & Liu, Y. (2012). Traffic incident detection using multiple-kernel support vector machine. *Transportation Research Record*, *2324*(1), 44–52.

39. Deb Nath, R. P., Lee, H. J., Chowdhury, N. K., & Chang, J. W. (2010, September). Modified K-means clustering for travel time prediction based on historical traffic data. In *International Conference on Knowledge-Based and Intelligent Information and Engineering Systems* (pp. 511–521). Springer, Berlin, Heidelberg.

40. Huang, G. B., Zhu, Q. Y., & Siew, C. K. (2006). Extreme learning machine: Theory and applications. *Neurocomputing*, *70*(1–3), 489–501.

41. Zhu, Q. Y., Qin, A. K., Suganthan, P. N., & Huang, G. B. (2005). Evolutionary extreme learning machine. *Pattern Recognition*, *38*(10), 1759–1763.

42. Polikar, R. (2006). Ensemble based systems in decision making. *IEEE Circuits and Systems Magazine*, *6*(3), 21–45.

43. Zeynoddin, M., Bonakdari, H., Ebtehaj, I., Esmaeilbeiki, F., Gharabaghi, B., & Haghi, D. Z. (2019). A reliable linear stochastic daily soil temperature forecast model. *Soil and Tillage Research*, *189*, 73–87.

44. Sudhakar Yadav, N., Reddy, B. Eswara, & Srinivasa, K. G. (2018). Cloud-based healthcare monitoring system using Storm and Kafka. In *Towards extensible and adaptable methods in computing* (pp. 99–106). Springer, Singapore.

45. Yadav, N. Sudhakar, Reddy, B. Eswara, & Srinivasa, K. G. (2018). An efficient sensor integrated model for hosting real-time data monitoring applications on cloud. *International Journal of Autonomic Computing*, *3*(1), 18–33.

46. Wantmure, R., & Dhanawade, M. (2016). Use of Internet of Things for building smart cities in India. *NCRD's Technical Review: e-Journal*, *2*(2), 1–6.

7

Recent Advancement in Emergency Vehicle Communication System Using IoT

Ajay Sudhir Bale

New Horizon College of Engineering, Bengaluru, India

Vinay Narayanaswamy, Varun Yogi Shanthakumar, Parinitha Balraj Shyla, Shivani Balakrishna, Varsha Shyagathur Nagaraja, Sahana Basatteppa Menasinakai, Manish Kumar Hitesh and Eshwar Esarapu

CMR University, Bengaluru, India

CONTENTS

7.1 Introduction 121
7.2 Comparative Study...122
    7.2.1 Traffic Monitoring..122
    7.2.2 Patient Monitoring...126
    7.2.3 Vehicle Monitoring 130
7.3 Challenges of Smart City Applications...141
7.4 Summary of the Survey 144
7.5 Proposed Methodology...148
7.6 Conclusion 150
References..150

7.1 Introduction

In densely populated cities like Mumbai in India, traffic management is a mammoth issue where there are vehicles moving continuously making it tough for emergency vehicles (EV) to reach the intended destination which may result in casualties [1–5]. EV drivers find it challenging to reach the hospital in time due to lack of access to traffic related information which is on the way to the hospital creating delay in reaching the hospital risking the patient's life which makes traffic management crucial when seen from an EV point of view. Road traffic accidents claim the lives of about 150,000 people in India every year. Time is lost owing to delays in the arrival of an emergency vehicle, such as an ambulance, at the scene of the accident, as well as delays in the care of accident patients in the hospital. The Internet of Things (IoT) may connect to an ambulance by transforming it into a smart ambulance that can gather and send the injured person's health condition to a nearby hospital over the internet. This implies that before the sufferer gets to the hospital, doctors may assess his or her physical state and determine whether the victim's situation is indeed

DOI: 10.1201/9781003217404-7

serious. Therefore there is a necessity of a system which overcomes the said problems and provides fast and secure services to increase the probability of the casualty's survival. This work proposes a method in which IoT helps in the traffic management so that the nearby hospitals can be contacted in getting quick assistance [6, 7].

7.2 Comparative Study

7.2.1 Traffic Monitoring

To deal with the problem of traffic management [8], we have Vehicular Ad Hoc Networks (VANETs) [9]. VANET consists of two categories which are Vehicle to Vehicle Communication (V2V) and Vehicle to Infrastructure (V2I) technologies [10] which create communication to manage traffic automatically and assist EV to choose a route with least congestion on roads and help them to reach the destination in time. According to [11], VANETs have a distinctive traffic thickness which concludes that the network probability models can meet the true circumstances of VANETs. In [12] an analysis on VANETs clustering techniques in detecting and finding the traffic congestion on roads with the least assistance from infrastructures is proposed. The centroid-based K-means algorithm [13], Object-based Fuzzy C Means (FCM) algorithm [14] and the Fuzzy K Means (FKM) algorithm [15] are used. The data generated by the vehicles are distributed in various parameters using these algorithms. When an EV driver wants to know the traffic congestion, an On Board Unit (OBU) in the vehicle transmits a route status check message to the Road Side Unit (RSU) to collect the route data such as traffic congestion, accidents etc. for smoother travelling. The RSU then sends back all the acquired data to the OBU which then processes the received data and proceeds to route selection based on the data received from the RSU. The OBU selects the route with least congestion and plans the travel accordingly, helping the EV to reach the destination on time [16]. The Figure 7.1 shows an IoT-based vehicle communication system [17].

The technology utilised to enhance the quality of life is critical to the smart city concept's long-term survival and intelligence. As a result of the IoT, smart cities have evolved. Given that overcrowding is a major problem that worsens as cities grow, smart traffic infrastructure is a critical sector of smart city projects. Smart health care, building management, traffic management, and parking solutions are also important [18–22].

Maps correctly anticipate congestion problems for metropolitan roadway based on the data retrieved from surveillance equipment installed on highways or city roads. To collect traffic data, these app developers build agreements with a variety of transportation agencies [22]. They do expect smart car technology, as well as any smart cellular device in the hands of the driver, to provide real-time driving directions. Drivers want to know how congested future intersections are so that they may plan accordingly and save duration on the road by using substitute routes.

The proposed system in [23] consists of two modules: the first one for vehicle tracking and the other for priority management. A collection of ultrasonic sensors are utilised to power it. The system identifies any improper vehicle behaviour, such as jumping traffic signals, inclusion to traffic signal lighting. Despite the fact that no smart devices are planned to be utilised by drivers in this study, updates regarding traffic via roadside message units are carefully investigated. The portable gadgets are primarily used to warn vehicles of potential traffic jams. The majority of the notifications on the roadside units

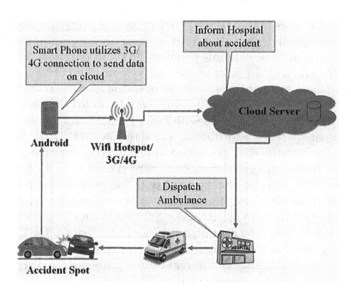

FIGURE 7.1
IoT in emergency vehicles [17]. Reproduced from an open access article.

are about overflowing roadways, planned activities, weather information, road traffic conditions, and other similar topics. A technique of displaying traffic intensity using three separate light colours is recommended on mounted electronic boards at the sites of decision. In this approach, the actual traffic load is obtained using the mean vehicle speed collected from vehicle detection sensors [20, 24].

Magnetic sensors help in the perception of the disruption in the earth's atmosphere caused by traveling automobiles and quantify it as vehicle magnetic length (VML) [16], which is needed to compute the vehicle's length. A ferromagnetic material sensor can accurately identify an automobile with 99% precision [25], a few sensors can estimate speed, and more than two magnetic sensors can categorise a vehicle, according to the research. Real-time and notifications on uncommon road occurrences are the two forms of traffic warning messages. Drivers may get these updates via smart mobile applications, smart phones, iOS, TVs, and other devices. Another option is to place boards on the side of the road at major junctions. These units will assist the greatest number of people and will assist them in taking other routes [26, 27]. This finding suggests a system structure based on the IoT for collecting, processing, and storing real-time traffic data. The traffic administration can transmit priority notifications to residents, avoiding traffic congestion caused by accidents or other unusual events. The prototype in [18] demonstrated high vehicle identification accuracy and a very less error in estimating the occupancy of the vehicle on the road.

Traffic congestion usually depends on various factors such as traffic density, natural disasters such as earthquakes, fire accidents, sometimes due to civilian incidents such as human strikes, road accidents and in other cases due to government related construction activities. In [28], a research by the United States Department of Transportation that monitors road networks and the relevance of various escape routes that emergency vehicles can choose from to reach the intended destination is presented. According to this report, traffic congestion is classified into two types, planned and unplanned events. Planned events involve road constructions and road closures whereas unplanned events involve emergency events, weather incidents, and natural disasters. In these unprecedented situations, getting an EV (example, Fire Engine) to the destination is crucial. To tackle this problem, IoT and data fusion techniques pose to be promising for efficient traffic management systems [29]. This facilitates guiding an EV (e.g. Ambulance) to easily cross the urban traffic and reach the hospital in time, increasing the probability of the casualty's survival. In an IoT architecture is proposed which avoids road blockages and roads with more traffic. Loop monitors, wayside sensors, airborne sensor systems, and floating car data (FCD) [30] are only a few of the methods for collecting traffic data described in [31]. These techniques are involved in gathering the data from different automobiles which consist of an assortment of sensory devices. This FCD data can also be used with the proposed IoT system to obtain traffic data. Apart from obtaining traffic data, it is important that an EV should be routed with an alternate route to reach the destination in time which can be done using a routing machine or engine. The routing machines [32, 33] assist the EV's IoT architecture in determining the shortest possible route to the destination which is a hospital in case of an ambulance. The traffic data in real time is gathered from sensors where few sensors consisting of GPS unit is connected to a Microcontroller Unit (MCU) which is Wi-Fi enabled constituting a geo sensor node [34]. The sensor is tuned to create a Wi-Fi signal with the nearest accessible access point, and all data obtained from the surroundings is transmitted to a server by the Traffic Monitoring Control Centre (TMC) [35]. The geo-tagged sensor data and crowd source information [36] are stored in the server of a database on a web server at TMC. There are two stages to implementing a decision support system. The data is aggregated and analysed in the first stage using a fuzzy logic-based data fusion approach. It then aids the routing engine (i.e., OSRM) in producing aware of congestion in the following step. When the OSRM server receives a route entreaty for an expected hospital in the event of a paramedics, it sends a route reply to the user.

To tackle the problem of delay in arrival of an emergency vehicle to the hospital the system in [37] is divided into two sections: an ambulance and a hospital. Many health monitoring devices are installed on stretchers in ambulances to monitor the patient's live health data. Every second, the server sends this information to the hospital. It's also important to notify the local police station about the accident and the patient's personal information. To obtain patient personal information, a biometric sensor is employed. An alert will be sent to the patient's family using this information. One portion is dedicated to ambulances, and the other is dedicated to traffic. The Atmega 2560, a microcontroller, was utilised in the ambulance segment. There are 54 analogues and 16 digital pins on the microcontroller. They use a GSM 800 model to send an emergency message to the hospital, family, and a police station close by. A traffic light's microcontroller can be linked to an ambulance's RF receiver. The RF transmitter in the vehicle delivers the signal to the RF receiver once it has reached 500 meters before the traffic light. Traffic lights can be turned on and off based on the signal, avoiding traffic congestion [37]. When a vehicle is involved in an accident, biometric sensors are used to identify the victims, and information is relayed to the victim's family and the hospital. The ambulance's path is cleared, and the patient's status is watched. The information is updated on a website that the medical personal may access so that appropriate discussions can be made. The Figure 7.2 shows the Management of Traffic using IoT [36].

Congestion in metropolitan areas has a detrimental influence on the city's sustainability [38]. It not only affects the functioning of the city but also affect EV in case of an emergency situation which may be a fire accident or a road accident where an

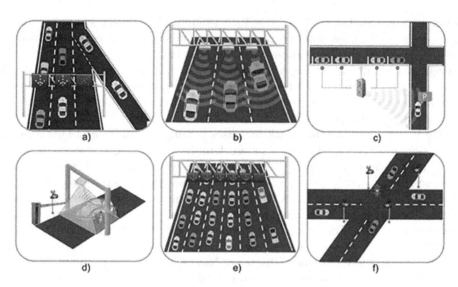

FIGURE 7.2
Management of traffic using IoT [36]. Reproduced from an open access article. (a) Lane management, (b) surveillance, (c) parking management, (d) automatic tolling, (e) special event transportation and (f) intersection management.

individual is involved, getting the EV to the event site and back to the destination which may be a hospital in case of an ambulance is a huge challenge [39]. To deal with such tough situations, we need an effective congestion prediction system which helps EV to reach the destination in time and do the needful [40]. However, predicting congestion is itself a challenge due to the changing behaviour of the traffic. The data that is received is then analysed to predict the traffic congestion in the near future and assist EVs in reaching the destination in time. This collected information is utilised in providing the traffic pattern to users so that they can plan their travel accordingly. The work in [41] presents a congestion prediction system that involves IoT and Long Short Term Memory Models (LSTMs) [42] to predict traffic congestion based on the detected or predicted speed of a vehicle in a duration of 5 mins within a town. These models can provide reasonable traffic congestion data over a short period of time with an accuracy of 84 to 95% with the 5mins time slot. The various sensors which are mounted on the roadside [43] and the vehicle transmit the data to the server which is then analysed by the LSTM models to predict congestion in the near future and helping users and EV in reaching their destination in time.

An unlucky accident on the road or a roadside repair work might cause a major traffic jam and subsequent delays in vehicle movement. Traffic signals are crucial in reducing traffic flow disruptions and delays. It analyses the generated data to help in traffic management by utilising social relationships among the entities and interactions between them. When the number of cars on the road exceeds the capacity of the road, it is known as traffic congestion. The work in [44] presented a technique for detecting and quantifying traffic congestion and proposed a continuous-time route reservation architecture for managing route to the vehicles and reducing congestion [45].

Social IoT (SIoT) is defined as a new IoT paradigm in which objects are capable of forming social interactions with one another. It has presented a prediction model that captures spatio-temporal characteristics and a wide range of co-usage data to infer the social interaction between SIoT items [46]. To reduce accidents, [47] outlines how to establish an effective exchange of congestion information and by raising awareness on the road and in the surrounding area, the distribution system may be improved.

To reach the objective of having a good road traffic that is clear of jams and incidents, different connections can be used to automobiles. Experiments with the 802.11p protocol are difficult due to the limited number of suitable items on the market [48–53].

The work in [50] introduced the Dynamic Throughput Maximization System (D-TMF), a throughput optimisation framework that uses turning intentions and lane locations to enhance traffic flow. The junction road network motion scheduling creates a binary phase matrix, which displays the non-conflicting allowed movement kinds. It presents a method for dynamically organising non-conflicting traffic to maximise its throughput. It creates a road junction with numerous lanes of distinct vehicle movement types on each route. While travelling along the road, the cars form social bonds with one another depending on their requirements [51]. The work in [51] explains how to distinguish between EV and regular cars using several approaches. It also offers various methods for dealing with traffic for both regular and emergency situations. New solutions for coping with traffic congestion at junctions are developed on a frequent basis. The Figure 7.3 shows Management of Traffic Signals Using IoT [26].

7.2.2 Patient Monitoring

By 2022, IoT participation [52] in the health business will be approximately $409.9 billion. Smart phones, biosensors, wearable gadgets etc. are ushering in a new age in health care. The general features of mobile/wearable sensor systems helps in monitoring the safe function of the heart. Body Sensor Network (BSN) nodes [53], CPU, central web server, patient database, and a PC are the five essential components of the proposed system in [54]. Wireless electrocardiogram (WECG) [55] and pulse oximeter (SpO2) sensors are included,

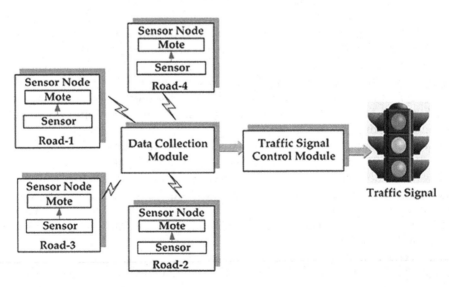

FIGURE 7.3
Management of traffic signals using IoT [26]. Reproduced from an open access article.

as well as other more generic sensors. The IoT helps the smart devices to establish a means of network by which data can be gathered without human involvement. In 2008, the number of devices assigned IP addresses outnumbers the global population. The way emergency and paramedic services function is changing as a result of IoT technologies [56]. If something goes wrong with a patient, important information can be quickly forwarded to the intelligent systems in quick time and the medical personal can be quickly notified about the situation [56, 57].

Among the most transparent and commonly utilised IoT-enabled medical services is telehealth. Patients may not need to attend the emergency department or even the hospital in some circumstances thanks to the IoT. This not only saves money and reduces the need for the user to visit the hospital, but also reducing the burden of having to travel to the hospital enhances the adherence to treatment [58].

With longer lifespans and an increasing senior population, present healthcare systems will clearly be unable to meet the demands of older patients. The internet of different items enables for continuous monitoring of older people's health data [59, 60]. This will make life much simpler for those with Alzheimer's, Parkinson's, and Dementia, as well as their caregivers and family.

Wearable technology, next-generation analytical equipment, and mobile devices can all help to provide new ways of diagnosing and treating chronic illnesses [56]. IoT devices are used in Fitbit-like applications to enable continuous long-term health monitoring. Health Net's diabetes management program seeks to enhance diabetic patients' clinical care while also lowering their medical expenditures.

In the health industry, IoT applications provide new custom prescriptions and medicines. Personalised medicine can transmit signals to an extrinsic device connected to the patient in order to ensure right dosage and drug consumption. Patients will now be able to track their progress and make adjustments using a smartphone app that will give them access to all of their personal information.

Many medical companies across the world are moving to IoT and real-time location solutions to make resource and service tracking easier. The goal is to provide order to these enormously intricate things in hospitals, such as staff, equipment, and safety. This implies that while monitoring hospital resources such as employees, patients, and equipment is simple for local facilities, it is a nightmare for major city and metropolitan hospitals.

The work in [54] discusses a prototype health information monitoring system based on ZigBee. The patient's pulse rate, plethysmogram, and blood volume are electronically sent to a data base. A limit for each patient may be determined if necessary, and a display alert or other alerts might be sent to a control centre [61–63].

The significance of Golden Hour in traffic accident situations is self-evident. The suggested system will locate the closest hospital, determine the shortest route, communicate critical parameters to the hospital ahead of time, and regulate traffic signals to ensure the ambulance's safe passage. An efficient and successful method of assisting in the achievement of the objective of preserving a life. The Figure 7.4 shows the Patient Health Monitoring System [63].

In today's fast paced lifestyle, people are not concentrating on their health making them prone to germs and diseases. It's critical to have an eye on the health of an individual which can't be monitored continuously by doctors and hospitals due to the huge population. The growing demand for both health monitoring and technology [64, 65], has resulted in using IoT in monitoring health. It's a new age technology that can be used in many applications to transmit and analyse data and come up with solutions. Healthcare is one of

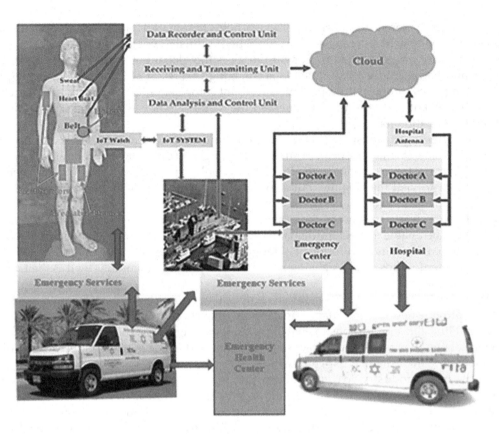

FIGURE 7.4
Patient health monitoring system [63]. Reproduced from an open access article.

the most common popular applications of IoT and is growing exponentially. IoT helps in acquiring valuable health data such as heart rate, oxygen levels, pulse, glucose levels using biosensors and transmitting these collected data to the central server where the doctor or the hospital can have access to this data [65–67]. To collect this data, the sensors need to be connected to a network where Wireless Body Area Networks (WBANs) come into picture. Communicating the patient's health data is of paramount importance. Various Patient Health Portals (PHP) are being developed as they are crucial in the remote healthcare link. Using this facility, patients can contact the doctor or the hospital for routine diagnostics and bill payments. In [66], a unit sends ECG information to the server using a Zigbee network. Thanks to research, now sensors can be situated on the body of the patient which sends various health data wirelessly without the user experiencing any pain while doing his/her daily chores. This application can also be deployed in emergency vehicles where a suite made of various biosensors can be put on the patient in the vehicle and this system wirelessly transmit data using WBANs [67] to the central server and the doctors and the hospital can get ready with all the arrangements need to be made before the patient arrives to the hospital thus cutting down the time required to make all the necessary arrangements which in turn increases the patient's survival and also makes the hospital to respond swiftly. The main benefits of WBANs is that it can provide health services outside the hospital complex thus enabling remote health services helping reduce the workload on

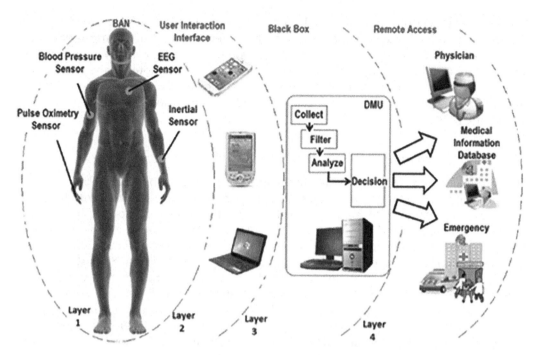

FIGURE 7.5
WBAN-based IoT health monitoring system [68]. Reproduced from an open access article.

hospitals and making them work more swiftly [68]. The Figure 7.5 shows the WBAN implanted in a body for communicating with devices [68].

The Covid-19 illness has spread to more than 200 nations, causing a pandemic. This virus is part of a large family of viruses that may infect both animals and humans. People can be affected by inhaling droplets from a virus-infected person. These droplets can adhere to nearby objects or surfaces, such as tables, doorknobs, and handrails. When people come into touch with the sufferer, the surface, or items, they become contaminated. In this pandemic condition, it is recommended that you keep at least one meters apart from people [69–71].

The IoT application offers a lot of benefits that make it a top choice in the medical industry. Due to IoT connection, medical staff can monitor Covid-19 patients and those who self-quarantine from afar. This means that medical staff will be able to get all of the information they need about these patients in one place and decide what to do next. The applications will effectively limit person-to-person interaction, lowering the Covid-19's long-term spread [72, 73]. Data uploaded to the cloud is encrypted. In an emergency, this information is useful for deciding on a patient's concern.

IoT provides a smart dedicated gateway to combat the Covid-19 Pandemic [74]. Healthcare monitoring devices are connected to the internet thanks to this high interconnectivity [75]. It may send a message to the physicians at the hospital/healthcare facility remotely in any emergency case. All of the data is remotely monitored at the healthcare facility, and numerous statistical analyses may be performed to anticipate Covid 19's future situations.

IoT will have a significant impact on driving behaviour in order to assist prevent accidents, resulting in increased safety and a potential reduction in occurrences.

However, a substantial number of IoT-enabled cars will be part of controlled fleets in the automotive industry [76], including EV that offer vital and lifesaving services for patients as well as assistance for medical personnel. Individual cars will be watched and tracked, route planning will be more efficient, and even remote repairs will be possible thanks to the IoT and linked automobiles. All of this will result in time savings, cost savings, and less downtime for vehicles whose operation is vital to preserving human life [77].

7.2.3 Vehicle Monitoring

The accidents are caused majorly due to the improper driving operation of vehicle drivers by rash driving, alcohol driving, fast turn, high change of speed along with this road traffic are more difficult to manage vehicles using traditional methods [77]. To avoid this IoT technology came up with many ideas to resolve this issue and the research is still going on to implement better methods. Many vehicles are designed with sensors to monitor engine operation, speed of vehicle and other operations [78]. A study refers to vehicle behaviour analysis using IoT that determines the condition of a vehicle, efficiency of running speed and confirms safe operation to avoid the majority of traffic accidents. In this system, the architecture design contains three layers namely perception, transmission and application layer along with three core parts. First core part is the vehicle data acquisition terminal which gives the update of vehicle parameters such as operation, motor and driving. Second core part is the 3G wireless transmission network used to establish connections. And the third core part is the remote monitoring centre server used for communication. Here the perception layer collects the data of vehicle operation and is then forwarded to the centre server which monitors remotely. Finally in the application layer that specific data is displayed on the PC server of monitoring software. To analyse the purpose of storage interaction, terminal of vehicle, vehicle operation analysis and interaction with clients the data centre is constructed using Alibaba cloud and Hadoop [79]. The vehicle behaviour monitoring system is constructed using the Internet of Things (IoT) and then the Hadoop data network model is established. Finally, the expected result is received after comparing the simulation results with the manual traffic data.

The automation industry is increasing rapidly to develop smart vehicles. In the modern world to make a smart vehicle monitoring system an in-built ECUs and many different sensors are used to establish [80]. Along with this OBD2 is added to get detailed data to identify the issues occurred in the vehicle operation [81]. Online vehicle health monitoring systems use different sensors which are fitted onto the vehicles for fault predictions and engine condition can be detected using machine learning algorithms. Mainly diagnostic and remote prognostic of vehicles is provided using three layered IOT stage architecture which is a raspberry pi-based system that uses random forest for fault detection. This system depends on the OBD2 (On Board Diagnostic-2) scanner as the input. It detects the fault in the vehicle and passes the information to the driver by flashing Malfunction Indicator Light (MIL). A hand-held scan tool is attached to the Data Linker Connector which is present at the bottom of the dashboard to display the information through CAN bus interface after communicating through OBD2 using trouble codes. Hand-held scan tool provides information in mechanical terms but it does not mention time while displaying the details and it is difficult to understand. Along with a warning alert, it provides suitable measures to deal with the issue and even shows mechanics shops nearby in the mobile application. This system mainly concentrates on the ignition, fuel, exhaust and cooling system of vehicles [82]. The idea can be implemented further to check the consciousness of the driver using cameras, tracking vehicles by GPS module and preventing accidents using ultrasonic sensors.

In the smart vehicle system, OBD-II is used in addition with ECUs and the different sensors to find out the issues of operation and to understand the deep data network. Here the experiment is conducted using On-Board Diagnosis (OBD-II) which is low cost, easily installable, and simple to access and transmit the data. The idea is to check the engine's condition and to monitor accurate real-time weather conditions using IoT wireless sensing [83]. OBD-II data acquisition can be performed only in wired or offline mode. To overcome this the experiment is performed. The system consists of three main stages. Stage one is CAN-bus data collection from the Electric Control Unit (ECU) using two methods. The first method is done using a transceiver TJA1050 which collects data and sends it to the microcontroller through serial communication but it has a data clocking problem because without segregating it sends packets of data to the microcontroller. The second method is using MCP2515 and TJA1050 CAN data converter board which can segregate each byte without messing up and also it helps in selection and conversion of data packet. The data is converted into decimal for human understanding in the second stage. And finally, stage three is sending data to the cloud in which the converted CAN-bus data is stored using two methods that is HTTP request with cloud data.sparkfun.com and Fusion Tables with google cloud storage. Figure 7.6 shows the flow of engines diagnostic using IoT. After experimenting with the system platform, the hardware implementation setup is constructed. The ECU transmits data packets which consist of CAN-id. This connects Arduino through SPI and CAN High/Low. Arduino segregates packets of data to get the raw data of the engine. Obtained data from ECU is sent to the cloud through ESP8266 Wi-Fi module in architecture REST method. Simulation results were obtained in two methods [84] but the best result was obtained using MCP2515 and CAN-bus converter board TJA1050.

One more study is based on J1939 protocol monitoring the special vehicles condition with the help of Electronic Control Unit (ECU). For standardisation of buses this protocol has been implemented and also to increase the communication speed between ECUs [85]. Based on the Open System Interconnection (OSI), J1939 protocol has been built and each layer is described with corresponding documents, including the layers namely data link layer, vehicle application layer and diagnostic application layer. Data link layer can explain all the features of the

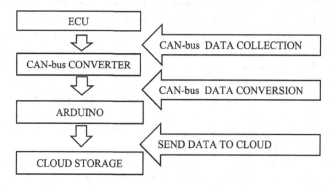

FIGURE 7.6
Flow for engine diagnostics using Internet of Things [83]. Reproduced from an open access article.

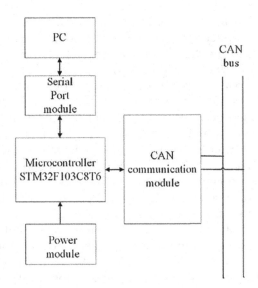

FIGURE 7.7
Architecture design flow [87]. Reproduced from an open access article.

message because it redefines the identifier of 29-bit. There are two types of application layers in this protocol [86]. Information related to the vehicle parameters and communication is defined by the vehicle application layer. Diagnostic message and diagnostic trouble code is defined by the diagnostic application layer. The architecture design consists of a microcontroller (STM32F103C8T6), CAN communication module, serial port module and power module as shown in Figure 7.7. The power supply to the microcontroller, CAN communication module and serial port module can be given when the power module depressurises the battery. CAN bus collects all the data from ECU and forwards that data to the CAN communication module. This communication uploads the data to the microcontroller. According to the J1939 protocol, the microcontroller performs a message and uploads the fault information and parameters related to the operation to the PC for displaying through the serial port module. At the same time, through the serial port PC can also send request, which will be processed by the microcontroller and uploaded to the CAN bus, and then wait for the ECU response. The required output has been obtained after testing the implementation [87].

Mainly in countries like India, the unmaintained roads are more. Especially during the rainy season the roads become slippery. Due to that slippery slope it is difficult to drive vehicles which leads to accidents [88]. Car monitoring using IoT focuses on increasing the efficiency of the vehicle and also to get the monitored report for the daily maintenance of vehicles. IBM Blue-Mix is a cloud computing platform which can be easily used for mobile applications. OBDII is a scanner tool which can be communicated with other applications through IBM Blue-Mix. In this system the car Electric Control Unit based on OBD parameter ID such as location and other information of PID is sent to the smartphone through Bluetooth. The received data to the smart phone is forwarded to the IBM Blue-Mix which stores data in the database and that data is displayed in the web pages to analyse the vehicle condition so that the user can access it for better vehicle management. Along with this speed of vehicle status can be seen when there is variation in the speed of the car. After implementation it was tested using a car manufactured by Toyota [89].

The major reason for the cause of road accidents is due to behaviour of the driver while driving. One of the solution is to create awareness about driver safety [90]. Smart

Phone-based condition monitoring of vehicles is developing but very little effort has been put on the vehicles which we ride daily due the condition monitoring equipment cost. To overcome this problem, a cost-effective based condition monitoring vibration unit and a technique of Enhanced Sampling Rate (ESR) for collecting data is proposed. ESR technique is used because smartphones have the limited capacity to capture images. This method is performed to know the device limitations in the form of the sampling rate. Nyquist criteria is a key requirement according to the fundamental sampling rate. Analysing the high frequencies of resonant band frequency is desirable for the purpose of the car monitoring system [91]. This technique gives a new dimension for monitoring the condition of a vehicle, fault diagnosis and prognosis.

The PATH research, which defines varying aspects of command and control functions utilising five separate operational layers: network layer, connection layer, link layer, control layer, and body layer [92], led to the development of Advanced Vehicle Control Systems (AVCS's) well-designed control structure. Each operating layer performs among the roles listed: movement of traffic, rush hour management, vehicle traffic control, command control, and hearing.

In the designed structure, this sequence design enables the distribution of inefficient management. In most cases, these devices are built into the application build or part of it. The safety of AVCS was judged on vehicle safety [93]. The speed of the car can be adjusted automatically to maintain the distance between it and the front car or above the specified small safe minimum distance (d_{min}) and adapt to the acceleration and deceleration of the front car. The other works in this field have recommended having a link in between the two automobiles so that traffic congestion is reduced. The long-distance sensor simply monitors the flow in this fashion, and the transceiver allows the following car to get velocity information from the vehicle in front of it. Although the sensor used may provide support ratings, the next vehicle may not be able to avoid certain hazards as the small headway is designed to make extensive use of the communication link.

Failure to detect and assess risk is two important functions of a health monitoring system. The following are examples of health monitoring strategies: The head control system achieves full efficiency during normal mode operation by inserting a secure connection link, as shown in this example. Sensor Discovery: Error detection can be achieved by explicitly installing hearing aids to ensure modest but safe operation if the communication link is broken. Hearing devices may transmit a restricted signal to the sensor control system or other sensitive security system if the visual module's performance falls outside the defined range. A separate sensor process is used on each sensor device to provide accurate comparative findings, as the source of failure is likely to cause erroneous events in all sensors. Methods using models can be used to measure exclusive of the output and to measure operational accuracy.

The purpose of the work in [94] is to examine the health of motor vehicles and how they relate to the management of malfunctions. It oversees the requirements and functions of vehicle testing, performance monitoring, and failure detection and diagnosis. Using the failure mode results and critical analysis, it determines AVCS critical security functions and key elements that make critical security functions. Health monitoring is an important function of the program or a critical part of security, according to the findings. If the performance of a program or part cannot fully be considered a comprehensive health monitoring system, it cannot be considered a critical safety plan.

The Vehicle Health Management (IVHM) concept is the enhancement of diagnostic and prophylactic systems [95–97]. New and old vehicles must be equipped with complex technological functions that allow for the construction, operation, maintenance and informed support decisions [98]. This is entirely in line with the adoption of the IVHM concept which uses a strong combination of different practices from engineering science, computer science and communication technology to achieve cost-effective and efficient spending. Such studies require a detailed and accurate understanding of the available information.

The purpose of the research provided in [95] was to discover, translate and integrate current IVHM documents. Aside from the many potential benefits of IVHM, the difficulty of accurately balancing trading between associated costs, risks and rewards remains a major barrier to implementation. By connecting the care of new and old cars, IVHM has the potential to increase the safety, functionality and performance of current and future health care, but also reduces the cost and complexity of technology.

IVHM monitors and analyses this practice by combining methods, sub-methods and systemic approaches to provide the most comprehensive information and decision support in the performance of the entire automotive sector. This has been made possible by the latest advances in sensory, telecommunications, and software technologies, and seems to be an excellent opportunity to improve product management throughout its life, far beyond the automotive systems and perhaps even more sophisticated technologies. It covers property.

Conventional management techniques and concepts have reached their limitations as a result of the constant rise in vehicles and drivers, and are unable to satisfy the fast rising urban traffic needs. The Layer of sensing recognises significant things, transfers them from the real to the digital world, generates a digital representation of the size of the information for the car and driver, and enhances the vehicle and driver using a number of advanced automated identification techniques [99].

The sensory data in the IVHM system is divided into two types:

Offers a layer of application that manages the collected information resources, as well as functional APIs for display terminals, and radios to express the functions of the entire system. It can handle and service IVMS 'core items such as cars and drivers in the real world, resulting in a contemporary and intelligent transportation system focused on the automobile. One electronic credential belongs to one vehicle and the other to the driver, which we have abbreviated as vCard and dCard.

As the vehicle and its driver pass through a surveillance base station, the system receives vehicle and driver identification information as well as traffic data, and accesses the vehicle using radio and creates information resources about its drivers. Frequency detection and video-perception technologies, VCard and DCard's ID number RFID antenna is automatically detected by the vehicle carrying them in the array detection zone.

The "Dual Card Matching" program processes and filters them.

- If abnormal data is found at the data collection base station, this passing vehicle will be recorded as an exception record.
- If abnormal data is found by the monitoring base station, the barrier is activated to block the vehicle and the location reaches an individual.
- It transfers the reason for the alert data to the information centre after recording it for the surveillance cell tower.
- The surveillance base station is responsible for controlling obstacles and traffic signals to keep the vehicle and driver under control.

The information centre monitors vehicles and drivers. The model in [100] is designed to record vehicle data and exchange it with various relevant businesses or segments to meet their shared needs using contemporary telecommunication technologies. According to the functional requirements and internal connections of the information service system, the collected data will be dug up for further useful data in classification, statistical evaluation, and integration and transport management decision making. It orders the cluster of base stations, controls the base station runtime status, and provides APIs for external services to enable data usage in multiple connected areas.

In response to the need for rigorous integration with grid control and public safety functions, more intelligent surveillance bases will be built on the city's main streets, using IoT that combines RFID and video recognition. Vehicle monitoring can be achieved scientifically, efficiently and in a coordinated manner.

Internet Data collects vehicle data as well as road network status in real time after the introduction of the Off-Things-Based Intelligent Vehicle Monitoring System. As a profound revolution in vehicle surveillance, it fulfils the demand for government and public control agencies, helps vehicle owners and drivers, and supports automotive management skills across the region [99].

The automobile industry has been growing rapidly since the beginning of the twentieth century. The evolution of unmanned vehicles increased dramatically in the early 2000s. In relation to this, it is important to monitor driving potential and save data on the vehicle's key systems in case of an emergency. In an emergency, it is important to monitor the driving operation and save information on the car key systems [100]. The complex is a "black box" that gathers information about instrument indicators and location, telling you what occurred before the event. Client-server design, in which the server contains data resources and if the server is the origin of these resources, computer system architecture is best suited for centralised algorithms. Both vertical and horizontal scaling [101] must be used.

Data synchronisation between nodes in a single layer is required by the three-layer concept of an analytical system distributed with horizontal scaling. Application servers seldom employ synchronisation, but allow server-side logic handlers to scale to independent nodes without sacrificing functionality. This architecture requires two-way communication so that jobs can be transferred from the application logic server to the multimedia data processing server. Fragments of the database layer are unloaded by direct communication between servers. Another example is that significant amounts of general data need to be retrieved frequently, although changes in this data are very rare.

The three-tier structure of horizontally scalable distributed computing systems requires data synchronisation between nodes in a single layer. Application servers rarely use synchronisation, but when they do, server-side logic allows their handlers to scale independent nodes without losing functionality. The three-tier structure requires two-way communication to transfer jobs from application logic server to multimedia information processing server. One can unload database layer pieces through direct communication between servers. Another example is a system that regularly accesses huge amounts of general data with very few modifications.

The researchers in [100] examined the server infrastructure and designed server applications that communicate with clients to manage data. Modern libraries and models are used in software. The libraries are popularly recognised, used extensively when building web servers, and are updated regularly. Flask, a popular contemporary library, was used to aid its development. Such a system is used in a variety of automotive sectors, including urban transportation, automobile rental businesses, and individual customers. The methods used are based on interaction with cloud technology, ensuring data storage stability and accuracy. Developed approaches can be used to create applications in various fields [100].

The convenience of digital infrastructure, one of which is the IoT, is a key factor in the adoption of Industry 4.0 (IoT). The IoT allows physical objects, cars, home appliances and other networks of electronic devices, software, sensors, actuators, and connections to link to the Internet and gather and share data. Automobiles must turn safely and comfortably. The frequency with which personal vehicles, especially automobiles, are used as a means of movement in daily life is expanding. To keep user performance and comfort at a high level, monitoring and maintenance are crucial. To do this, technology is needed, one of which is the use of IoT-based technologies [102].

This study [103] developed a prototype with an on-board diagnostic module and an embedded system that communicates data to the server in real time. The research aims to develop hardware for a prototype automobile vehicle position monitoring system that includes the IonBard diagnostic module that delivers data to the server in real time.

The system was created by combining a receiving device and a data processor to read machine data from an OBD device, as well as an RTC and GPS module to receive accurate data over time and space [103]. Data can be stored on a microSD memory card and sent to a database server via the GPRS module. Current data in the SD memory and database is analysed utilising real-time data visualisation as well as programs that allow extensive analysis to detect patterns of change and predict further engine performance. During this time, the machine was used to create a system model of diagnostic data reading used to create a description of system requirements in both hardware and software.

The on-board diagnostic unit, or OBD, is designed as part of the hardware module. OBD stands for On-Board Diagnostics, and refers to the ability to diagnose a vehicle system. Low-cost, low-power microcontroller systems are built with integrated Wi-Fi and dual Bluetooth modes. The communication module uses a full quad-band GSM/GPRS module with GPS technology for the purpose of satellite navigation [104, 105]. OBD (On-Board Diagnostic) is an automotive term that refers to a vehicle's ability to diagnose a system. The OBD system provides reports on the condition of their vehicles to car owners and technicians. The OBD is connected to the car ECU [102].

With the enhancement of metropolitan cities, challenges of traffic maintenance have increased due to the hike in number of vehicles. Due to which there is delay and disturbance when Emergency Vehicles (EV) are on the way to render help to the ones in need [106]. This calls for aid from the Traffic Management System and an improved, extended methodology that governs the traffic light duration based on dynamic aspects, along with the coherent management of EVs. It presents a solution to deal with deadlock (worst case) scenarios for vehicle supervision. Firstly, the condition of the patient in the vehicle is acknowledged by the nurse who possesses a gadget with GPS and Code Messaging Service through which location and patient's condition is updated to the Central Database. This information associated with a specific vehicle is then forwarded to all the points of access on the way to the hospital. The volume (queue length) of traffic, type and ID of vehicles are perceived and computed at each phase via Wireless Sensor Networks (WSN). These consist of inexpensive devices that process, sense and store information about the surroundings [107]. Later, the presence of vehicles at the intersections is monitored and the vehicles are supervised based on six classifications, in order to determine the best or worst case (starvation and deadlock) scenarios. Finally, the green light with an apt time span (<120 seconds) is allocated based on actively changing traffic constraints. With the frequent advent of EVs, the duration of green light is minimised so as to prevent scenarios of starvation. Various categories of emergency vehicles are designated with priorities based on standard guidelines and concurrent parameters. The procedure to control the signal in dynamic conditions is carried out using C++ with real-time [108] and Figure 7.8 represents

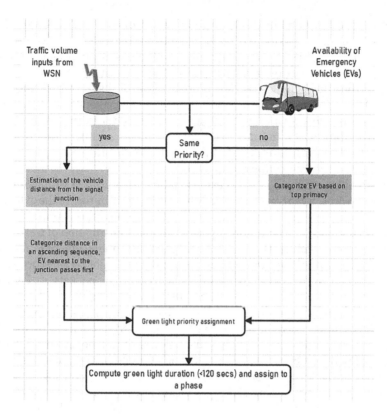

FIGURE 7.8
Smooth mobility of EV.

the scenario in [108]. The idea for this figure is derived from [108]. The methodology is an accomplishment from a theoretical point of view as it yields a solution to deadlock situations and also involves physical aspects that are dynamic. The complexity of time depends mainly on the Merge Sorting technique utilised to sort the vehicle distance from an intersection as well as the level of priority for each vehicle. This traffic management system provides an opportunity for further research and elimination of certain assumptions made during generation of control factors for the vehicle as well as the traffic signals.

Automated driving with safety system applications has been under spotlight for a while as safety is an essential factor for when a new mode of transportation is brought into play in real life. It is important for the smooth mobility and safety of vehicles, in particular emergency vehicles (EV) while reaching its destination. In order to make this happen, an approach based on automated driving mechanisms to develop a control tactic for EVs that focuses on an individual vehicle and also a preventive traffic signal approach has been discussed in [108]. A real time road section evaluation was carried out using a minuscule traffic simulation system to implement the proposed approach [109]. The EV is encountered with a GPS (Global Positioning System), V2X communication onboard unit (OBU) and sensors. The controllers for the traffic signal consist of roadside units (RSU) that facilitate retrieving wireless emanations [110]. For the lateral and longitudinal control of the vehicle, a selective lane switching decision algorithm with networks of signal processing and an Enhanced Intelligent Driver Module (EIDM) is utilised, respectively. A car-backing model, the EIDM is the source of implementation of adaptive cruise control (ACC) in vehicles [111] and ensures prevention of endangering activity while driving. In order to make the EV switch lanes under maximal manoeuvring capacity, the aggressiveness while driving is defined via the possibility of switching lanes and the probability of benefit in doing so. The priority and duration of green light signal, on receiving data from the vehicles, is determined by an EV signal pre-emption tactic to minimise any kind of delay and increase safety. The effectiveness of the approach was evaluated based on safety and mobility, both at network wide and individual EV level. Due to the consolidation of automated control of vehicles and the pre-emption tactics of traffic signals, the delay is minimal with the increase in mobility along with the lowering of vehicle conflict and enhancement in safety.

Industrial enhancement, geographical extension and urbanisation promote the amplification of traffic gridlock, accidents and even pollution, thus contributing to the exigency of traffic management. A fuzzy preference-based intelligent traffic congestion control (ITCC) scheme that improves accumulation, scheduling, organisation and transmission of traffic data in order to implement emergency vehicle (EV) governance and traffic management in ad-hoc networks is discussed in [112]. The primacy of road segment and green light signal schedule is resolved by a fuzzy proficient system and the dissemination of data packets to Dynamic Traffic Management Centre (DTMC) is achieved with minimal wait using a congestion-aware routing algorithm (CARA). Firstly, data is gathered and delivered via wireless sensor network (WSN) consisting of sensor nodes, then the accumulation of data and its channelling to DTMC is executed using slave road side units (SRSU) at the sides of each lane and a master RSU at the intersection. The DTMC inherits data from MSRU and initiates a fuzzy adept network to finalise priority and duration of green signal and road segments based on speed and distance of individual EV, as well as the volume of other vehicles. CARA, followed by priority emulation, schedules the packets for transference, then picks the forwarder candidate set by establishing clusters and impels weight-based relay intersection. This contributes to the utilisation of other possible routes to prevent and restrain congestion while forwarding data to DTMC with optimised link reliability and minimal delay. Low cost, facile deployment, scalability, simple maintenance, lucid operation, along with congestion curb, optimisation of EV travel duration and honed average waiting time of other vehicles are the accomplishments of the proposed system. Figure 7.9 is a representation of the basic operation of ITCC [112].

Technology pervades all facets of life which results in advancing countries espousing intelligent and smart traffic management, among which India is one of these countries intending to implement 100 smart cities consisting of digital libraries, smart schools, intelligent water and waste management and so on. The goal is to bestow an enhanced quality of life. An important aspect of a smart city is an IoT-based smart traffic control and governance system. This system can be executed via wireless communication grids, radio-frequency identification (RFID) and Worldwide Interoperability for Microwave Access (WiMAX). Certain methodologies using Green wave synchronisation [113, 114] and Zigbee transmitter were previously implemented but the former could lead to aggravated traffic if the synchronisation was disrupted while the latter has an operative range of only 20m and could not be applied in real-time traffic jams [115]. This led to the introduction of a novel approach named "Green Corridor" that comprises three various modules: a central controller server, RFID in the emergency vehicle (EV) and an RFID receiver at the traffic intersection. Figure 7.10 represents the Green Corridor setup described in [116]. WiMAX, made up of a tower and internal electronic paraphernalia, acts as the base station for the centralised controlling server and allows the EV to communicate with the server for best routes. WiMAX imparts a wide coverage area of up to 50km and an operating frequency of 2–11 GHz [117]. The server notifies all the traffic signals and the vehicle, thus including an asset of prior intelligence of the emerging vehicle that is equipped with an RFID

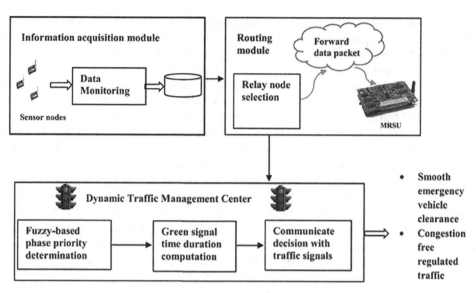

FIGURE 7.9
Operation of ITCC. Reprinted with copyright permission from [112].

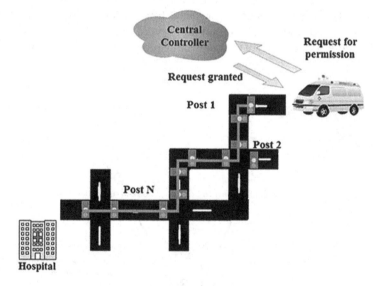

FIGURE 7.10
Green corridor setup. Reprinted with copyright permission from [118].

tag and an antenna. The RFID tag and receiver benefit in long distance detection of the vehicle [118]. On reaching an intersection, the antenna scans the RFID tag and the signal is set to green. And on passing it, the succeeding junction is alerted about the EV along with a timestamp using the RFID receiver. The upcoming intersections utilise the RFID reader to match the approaching EV ID and turn the signal to green on matching it, followed by the recommence of usual traffic. This technique was implemented using Arduino Uno microcontroller, a low frequency passive RFID reader and tag. The central server alerts the first signal post which in turn activates the green signal, triggers clearance mode of the EV and a notification with timestamp is directed to the NEXT signal. This procedure repeats until the vehicle reaches the destination and it also gives enough time to cease traffic, an advantage of improved performance, better efficiency and imparts Green Corridor of 1.2s to the EV.

Health care services are usually affected due to urbanisation and traffic jams, which is not only influenced by the volume of traffic but also due to post-hazardous circumstances such as natural calamities, strikes, fire accidents and so on. Governing the environment, emergency vehicles (EV) and traffic can be securely executed through conventional sensor networks [119]. A design using IoT integrated with multisensory data fusion [120] architecture was implemented to yield a system that guides EVs in smart cities, prevents road obstructions momentarily and avoids congested routes [121]. Utilising Raspberry Pi, tilt and sound sensors, an accident supervising framework based on fuzzy logic for the purpose of road accidents recognition has been accomplished in [122]. The guidance system for EVs makes use of sensors as the chief origin to acquire real-time data. A Geo-sensor node formed with a GPS device and particular sensors linked to a Micro Controller Unit (MCU) along with Wi-Fi, named NodeMCU, helps compute speed and emissions of the vehicles. The parameters mapped by the sensors such as carbon-di-oxide, sound, carbon-monoxide and temperature difference assist in the evaluation of the presence and intensity of the traffic. The speed of vehicles on road is indirectly proportional to the volume of traffic density hence, determining the average speed is essential. Laser detectors equipped with a sensor framework gather the speeds of vehicles and notify the MCU to determine the time lapse between lasers to yield vehicle speed. The nodes of sensors, with the aid of fuzzy logic accumulate, agglomerate and forward the data to the web server via Wi-Fi. This data is crowd sourced through an android ARM application to acquire information about the type, duration and severity of the traffic jam and the user can also perceive the route and location of the nearest hospital. Figure 7.11 is a representation of the

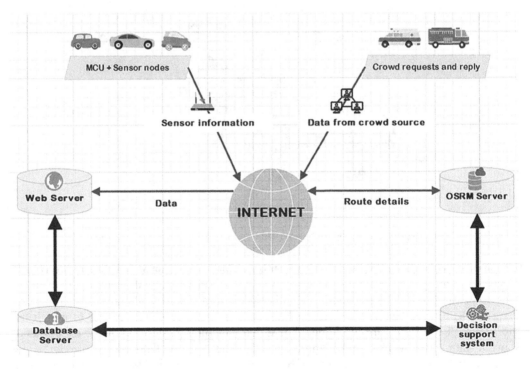

FIGURE 7.11
Architecture of smart city IoT framework.

framework of the Smart city based on IoT. The idea is taken from [33]. The genuineness of the user and data has to be verified by trust examination. An Open Source Routing Machine (OSMR), a high core routing tool [122, 123], is briefed by a decision making support system to bring about the minimised and congestion clued up route from the location of the user en-route to the destination. On asynchronous route appeal, about 4000 users are able to initiate requests with a CPU application of 38% and 2–113 ms of average delay in read, write operations. The lowest response time recorded is about 2 seconds with an error-free response when the amount of users is under 2000. The mean response size decrements with the increase in user and received bytes number. The recorded real time data can also be applied in various socio-economic circumstances like pollution maps, temperature prognostication and so on.

7.3 Challenges of Smart City Applications

IoT applications are made up of a collection of interconnected physical things, buildings, sensors, and electrical equipment that communicate over the Internet. In a nutshell, the IoT is everything that is linked to the Internet that collects or sends data. People may connect to the Internet through the IoT from anywhere on the planet. A smart city consists of a large number of sensors, engines, and, most importantly, people. The term "smart city" refers to the application of information and communication technology (ICT) to enhance a variety of urban services. ICTs are utilised in the Smart Municipal to enhance the consistency, performance, and interactivity of urban services, as well as to minimise costs and resource consumption, and to increase communication between people and city stakeholders.

The study's main goal [124] is to reduce energy waste and prevent injury.

- Public health is harmed by ultrasonic, LAN, and LoRa environments.
- Data and 6LoWPAN transfer from IoT sensor to Hamburg urban platform may be transparent and integrated by using a CoAP-to-HTTP proxy.
- Creating easy-to-maintain home gardens.
- ESP8266 gas cylinder temperature and level.
- Environment around the FC37 rain sensor.
- Monitoring temperature and watering in an urban greenhouse and garden (IDE), MQ135, agriculture in an urban greenhouse and garden.
- Introduce Fog Flow, a novel fog computing-based system strategy focused on design and development.
- NGSI and Data-Flow build a new fog (official standard) computing-based framework for a smart city.
- Sensor Type, which was a major issue in Sensor towns as well as on campus.
- The IoT setting.
- Individuals may use the IoT App to monitor and view ultrasonic.
- Traffic congestion is a typical problem in the system.
- HB, providing a Wi-Fi environment for employees, lowering the IR, and minimising the frequency of mortality at accelerometer construction sites.

The goal of the IoT is to build an intelligent world. It connects the natural world to the virtual world through smart items, data, and technology to smarten all urban businesses. It's a move that will change the globe by allowing for more networking. The IoT pattern is the centre of the ICTs section, which futurists and experts believe is the key driver of this transaction. Infrastructure-based data processing and dissemination involves smartphones, wireless sensors, cameras, and other devices that use networking technologies to process, gather, and transmit data via web providers. To make the Smart City concept a reality, several IoT deployment scenarios have been designed and created.

The IoT is a platform for connecting items through the internet to make system operation more efficient. Because the majority of data in today's world is communicated over the internet, IoT device security is essential. Citizens can perform things better and faster than before, thanks to Information and Communication Innovations (ICT) technologies such as embedded electronics and other sophisticated topologies in electrical sciences. There is an urban system that employs various ICT technologies to make infrastructure more interactive, efficient, and accessible than before; infrastructure should be accessible, and the demand for smart cities evolved as a result of a variety of factors. Because all of these varied infrastructures are made available to all different sorts of inhabitants, there are several IoT concerns in smart cities, such as security and privacy [125].

Considering all of these varied infrastructures are made available to all different sorts of inhabitants, there are several IoT concerns in smart cities, such as security and privacy. Citizens may be vulnerable to many forms of assaults as a result of this. Different tenants and users have access to the same equipment, and this multi-tenancy creates the danger of data leakage, data privacy leaks, data security threats, and so on. Due to vehicle movement, dependable and unreliable communication is not ideal. Device failures might occur, and this must be addressed. Large-scale deployment has its own set of issues [126]. The IoT is a fantastic concept for creating smart cities that are both efficient and beneficial. IoT can be deployed in smart cities, but only if all data security requirements are met, and issues such as devices consuming various levels of energy are scheduled and must perform duty cycles are addressed. In sensor networks, these are all distinct concerns.

Smart cities mainly focus on to give comfort services to main public resources. To communicate with everyone and everything, a new architecture 5G network has been introduced [127]. It allows a greater number of devices to connect and supports high data rate around 10 to 100 times than 4G network. 5G network provides improved privacy, security and consumes ten times less energy. This network can be used in industries like public transport, automotive, healthcare etc. To address challenges like transport systems, hospitals, waste management, etc. a new kind of connecting network is required. 5G network acts as accelerator and enabler for industries such as transport system, healthcare. 5G network in intelligent transport system is committed by linking vehicles individually to develop Cooperative Intelligent Transport System (CITS). When compared to existing networks, 5G is more efficient, provides safety and deals with issues like pollution, traffic congestion and collisions in the transport system. Now a day's industries and consumers are concerned more about safety, security and privacy as more devices are connected. The most major challenges faced are government regulations, cost, technological maturity and global standardisation. The reason because 5G network range is short, to increase connectivity range more vulnerable hardware is added [128].

Due to population increase, various open challenges are appearing all over the world such as energy, resource depletion and congestion in traffic, so many countries are trying to analyse these issues using IoT enabled smart city. Challenges that are presently being found as a prediction to make a success IoT enabled smart cities are overcome application silos, evolve towards flexible IoT platforms, business opportunities creation, achieve transparency and acceptance, integration on future networks, evolve towards multi-player, cross-organisation IoT platforms and applications and novel computing paradigms, develop resiliency to support mission critical applications, privacy regulations on data in different countries [129].

The world's population density is quickly increasing within metropolitan contexts in response to the need to meet citizens' daily needs. A smart city allows for the intelligent management of devices such as transportation, health, education, energy, residences and buildings, the environment, and infrastructure. It aids in the creation, dissemination, and promotion of sustainable development techniques in order to tackle the problems of expanding urbanisation around the world. The Smart City idea is predicated on billions of IoT devices being able to communicate with one another on a regular basis. The IoT refers to everything that is integrated with electronics, software, or sensors and can gather and exchange data through the internet [130]. The data sensed by sensors or actuators and transferred to the server through the gateway is the foundation of the IoT. Because of each company, organisation, and user, there is no one consensus on IoT design that is widely agreed upon.

The intelligent wireless metering scheme relies on two key components: the smart metering unit and the monitoring module. The sensor nodes collect data from the environment, which is then sent to the server via the MQTT protocol via the microcontroller unit. Because of its dual-core CPU, which can be locked at 80, 160, or 240 MHz, the ESP32 is employed in many IoT projects. With the introduction of IoT technologies, remote patient monitoring has gotten easier. This allows access to real-time information about the patient's physiological parameters, such as temperature and blood oxygen levels. After being acquired by the sensors, the data is collected in the microcontroller and then communicated to the server over Wi-Fi. The idea of smart cities is reaching its peak, and every country wants to be a part of it, especially in areas where information and communication technology plays a critical role. An application for smart cities was provided, which is using sensors to construct a smart monitoring system to monitor electrical energy use and monitor COVID-19 patients.

Smart cities employ a variety of technologies to improve citizens' health, housing, electricity, education, and water supply. Extensive data analysis is a comparatively recent technology that has the potential to expand smart city infrastructure. Successful exploitation and varied data uses are crucial in many industries and utility domains, including the intelligent urban domain. The IoT is a type of web application. Customers, items, and paper are all connected by interaction and information exchange. Many researches on the needs of creative and connected businesses have been conducted in this area. A Smart City should be able to pull historical and real-time data from a variety of sources in a variety of ways [131].

Mobility, governance, climate, people, and applications and services including healthcare, transportation, smart education, and power are all well-defined components of a smart city. Sustainability, longevity, management, higher living standards, and judicious use of natural and urban resources are all virtues. Big data has evolved into a strategic weapon with enormous potential, promoting industrial upgrade and growth as a crucial motivator for progress. Storage and computing bottlenecks obstruct traditional data processing systems as a result of massive, complicated, and variable data.

Big data mining is a service that gathers critical data and knowledge from a large data set and delivers it to the consumer. Instead of typical data mining, it helps to locate useful knowledge and expertise. However, there are technological, historical, data environment, and mining scope issues to consider. Cities and metropolitan areas are complex social ecosystems that include local government, inhabitants, and businesses. ICTs are increasingly supporting and enabling ICTs to meet certain requirements related to key issues such as enterprise and employment growth, economic development, energy and water, public security, the environment, health care, education, and government services. Simultaneously, the new chaotic global economic crisis is steadily putting downward pressure on urban spending, wreaking havoc on existing ICT infrastructure and facilities [132].

The term "smart city" refers to the use of applied research and technology, as well as information technology, to improve the infrastructure and service quality in the city. It has the potential to significantly improve and maximise urban management, services, and people's enjoyment of city life. Because of its exponential growth, the IoT has become intertwined with a wide range of fields and has infiltrated every part of our life. Smart city applications generate large amounts of data, which are subsequently analysed by big data networks to improve urban applications. Big data platforms can effectively archive, analyse, and obtain data from intelligent city applications, as well as collect data to improve various intelligent city services [132]. The benefits and problems of utilising large amounts of data in intelligent cities are discussed in this study. In order to implement the smart city definition effectively, efficient use and security of sensitive data is required. The study also looks at general Big-Data architectures for smart city apps and services, as well as basic design and deployment requirements.

7.4 Summary of the Survey

Vehicular Ad HOC Networks (VANETs) comprise of Vehicle to Infrastructure (V2I) and Vehicle to Vehicle (V2V) technologies that communicate to automatically manage traffic and help in emergency vehicle assistance to minimise congestion and delay [9–11]. Based on the VANET methodology, several techniques such as centroid-based K-means algorithm, Fuzzy C Means (FCM) algorithm and Fuzzy K Means (FKM) algorithm have been introduced. Gathering data regarding traffic congestion is done by means of an OBU that is present in the vehicle. This unit communicates with the RSU to extract data and then analyse the best possible route in order to reach the destination quicker, based on the data collected.

The IoT plays a significant role in enhancing the quality of life and bringing about change. Smart traffic management is one of the major contributions of IoT that aid in the reduction of delays and inconveniences caused due to traffic and accidents. In [23], the proposed model is based on priority management and tracking vehicles. In order to alert the drivers about congestion, weather or traffic information, portable gadgets are utilised and the mean vehicle speed is extracted from the detection sensors to estimate real time traffic load [20, 24]. [26, 27] have utilised IoT for gathering, processing and storing real-time information regarding traffic. Estimates of the occupancy of the vehicles on road have been executed with less error and higher accuracy. An analysis on road network monitoring has classified unusual events related to traffic into two categories, planned and unplanned events. In either case of such events, an EV reaching the destination is difficult yet crucial. One solution to this problem is the combination of IoT and data fusion approaches [29] that have shown significant efficiency in traffic management by guiding the vehicle to reach the destination in minimal duration. Wayside sensors, loop monitors, floating car data (FCD) [30] and airborne sensor systems are some of the IoT architectures introduced to gather data in order to avoid traffic congestion. Routing the EV towards a better route to the destination is a crucial step in avoiding any delay which can be successfully executed by routing machines.

Magnetic sensors can help compute the speed, length and even categorise the vehicle according to research [25]. Through smart devices or applications, real-time notifications and warning messages about unusual occurrences are updated to the driver. Sensors composed of GPS units are employed to gather real-time traffic data and passed on to a Microcontroller Unit (MCU) with geo sensor nodes and Wi-Fi [34]. The sensor creates a signal to the closest access points and obtains crowd source data which is transmitted by the Traffic Monitoring Control Centre (TMC) to store into a server [35]. Data fusion based on fuzzy logic is exploited to analyse the data to infer a decision and to bring about awareness of the congestion and the best route possible. Prediction of traffic congestion based on the detected speed of the vehicle was performed by a prediction system [42] including IoT and Long Short Term Memory Models (LSTMs). Traffic information with 84–95% accuracy over a short period of time can be retrieved by these models from the various sensors present on the roads and the vehicle. The information of which is then analysed for prediction. The ability to create social interactions is a recent inclusion to the methodologies introduced to avoid congestion and minimise accidents by the exchange of data and raising awareness. This is executed via a social IoT (SIoT) prediction model that obtains a wide range of data to infer interactions and analyse the spatio-temporal characteristics of the information.

Installation of health monitoring systems in EVs is required to not only constantly monitor the patient's condition but also to notify the patient's family, which can be executed using an Atmega 2560 microcontroller that employs a GSM 800 model to send alerts. The vehicle can transmit signals to the closest traffic signal microcontroller based on which the lights can be turned on or off in order to prevent congestion delay [35]. Prediction of the inconveniences or congestion related to traffic is an important yet challenging task due to the variations in traffic behaviour [40]. The data that is obtained from the situation is used to retrieve the pattern of traffic and then help the user predict the shortest and quickest route to the destination. The enhancement of traffic flow can be done by exploiting the turning intentions and locations of lanes via a Dynamic Throughput Maximization System (D-TMF) framework [50]. The throughput is maximised by dynamic organisation of the non-conflicting traffic that is displayed by the binary phase matrix from the junction motion scheduling. On a frequent basis, there are various methodologies being introduced to cope with traffic congestion and reduce the delay caused by the same.

Constant monitoring of one's health is crucial but cannot be easily performed due to the increasing demand and population which is why the entire medical industry is moving towards advanced technology provided by IoT. The application of IoT in healthcare is a significant and incremental aspect as smart devices, wearable gadgets and biosensors are becoming a herald for the new age of health line. The system in [64] employs a Body Sensor Network (BSN), central web server, CPU, database of the patient and a PC to help monitor a patient's health condition efficiently. A network of smart devices that require no human involvement are enough to gather crucial information, with the aid of IoT and has certainly changed the entire functionality of emergency and paramedic services [56]. Quick and effective help can be rendered to the ones in need by facile yet swift exchange of information. Telehealth minimises the need to visit hospitals and also saves time and money while adhering to the betterment of treatment [58]. Smart gadgets, wearable technology and next-gen analytical equipment lay new paths to the approaches of diagnosis and treatment of chronic illnesses. A personalised device connected to the patient and extrinsic equipment can corroborate drug consumption in the right dosage and also assist in tracking their improvement or deterioration in health. Real-time location solutions along with IoT provide advancement in resource and service tracking of vast intricate stuff like equipment, safety and staff, by the medical companies. Critical parameters such as pulse rate, blood volume and plethysmogram are monitored and transmitted to a database to alert the control centre in case of any complications, by employing a ZigBee-based health monitoring prototype [61, 64]. Health related information such as oxygen levels, pulse, heart rate, glucose levels and temperature are acquired via biosensors connected to Wireless Body Area Networks (WBANs) and are passed on to a central database which can be accessed by any medical help giver from any part of the world in order to render the necessary aid on time [58, 59]. Painless, non-contact sensors can be employed to wirelessly monitor a patient's condition and this approach can be exploited in emergency vehicles carrying a suit built of various biosensors that can constantly monitor and update the data through WBANs en route to the hospital [67]. This reduces the time required to personally analyse the patient and make necessary arrangements, increasing the chances of survival. IoT plays an important role in monitoring Covid-19 patients and also those who need aid during self-quarantine [72, 73]. This ensures safety of the doctors and the medical staff who render help during these tough times and also the treatment will limit to individual patients while bringing down the spread of the virus. In cases of emergency, the monitoring devices that are connected via the internet to the healthcare system provide a smart dedicated gateway to treat those in need while constantly monitoring the patient's condition for further aid [75]. This confidential information is encrypted and can be further utilised to analyse the patient's health. In the automation industry, a wide number of IoT-enabled vehicles will be employed to bring about a huge impact on the driver as well as the patient's safety [76]. IoT enables individual tracking of vehicles, remote services for repair and even efficient planning of routes, all of which will save time, money and even lives.

Malfunctioning of vehicles, driving under the influence of alcohol, rash driving, along with traffic problems create discomfort while managing vehicles through conventional techniques [77]. Sensors installed in vehicles help monitor the vehicle's functionality, speed and other aspects to confirm secure operation and prevent accidents or any other inconveniences. An architectural design consists of three layers along with three core parts. An information acquisition terminal for the vehicle, wireless 3G network for transmission and connection establishment and a remote monitoring server for communication are the core parts. IoT and Hadoop data network models are utilised to implement the vehicle monitoring system [79]. In-built ECUs, various sensors and On-Board Diagnosis (OBD-II) also play a significant role in vehicle monitoring and management. A raspberry-pi-based IoT architecture utilises three layers to remotely diagnose vehicles, detect faults and transmit this information along with a warning alert to the user via Malfunction Indicator Light (MIL). The best suited solution for the detected problem and the closest mechanic shop are notified to the driver. In another architecture based on IoT, data is collected via CAN-bus and through serial communication; this data is transmitted to the microcontroller unit. Data can also be acquired using the TJA1050 transceiver and MCP2515 through which the segregation of bytes and data packet selection can be performed. Data is then sent to the cloud and stored via HTTP requests and Fusion tables. Arduino is used to separate the data packets sent by the ECU which in turn passes the data to the cloud via ESP8266 Wi-FI module. The combination of CAN-bus converter and MCP2515 provided better performance and results. A J1939 protocol with ECU and Open System Interconnection (OSI) was implemented to monitor vehicle conditions and to optimise the speed of communication among the ECUs [85]. The architecture consisted of a data link, application and diagnostic layers. The data link layer redefines the identifier and describes the characteristics of the message while the application layer defines the parameters and communication related to the vehicles. The diagnostic application layer defines the message and trouble code. Again, CAN bus does the gathering of the information and passes onto the microcontroller which uploads the operational aspects and information related to fault to the PC [87]. The integration of OBD-II with IBM Blue-Mix, a cloud computing platform, is another IoT application that helps in vehicular monitoring efficiently. The ECU transfers data based on PID, location and speed to the smartphone via Bluetooth [89]. This information is then stored in the database of IBM Blue-Mix and then sent to web pages to diagnose the condition of the vehicle for enhanced management.

A control structure, Advanced Vehicle Control Systems (AVCS's) defines network, connection, link, control and body layers to gain command and control operations in management of vehicles [92]. The structure performs management of rush hour, facile traffic movement and control, hearing and command control. It enables the adjustment of the vehicle speed to ensure safe minimum distance and the adaptation of acceleration and deceleration. The vehicle receives data related to velocity while the system monitors the traffic flow. Complete efficiency of head control, ensuring safe and proper operation of the communication system along with the detection of faults are some essential requirements of an efficient health monitoring system.

An enhancement of the prophylactic and diagnostic systems is the Vehicle Health Management (IVHM) concept [97]. An amalgamation of computer science, engineering and communication technology has been utilised by the IVHM methodology to obtain efficient and cost effective measures in vehicular management. The problem of balancing trades among related costs, rewards and risks still poses a threat, despite the many advantages such as hike in performance, safety and functionality provided by this technique. The driver and vehicle related data are accessed and received by the base station for surveillance using the RFID antenna that detects the VCard and DCard ID [104]. Further processing is performed via dual card matching code which blocks the vehicle if any irregularities are detected in the data at the monitoring station and the location along with the reason for alert are passed on to the individual and cell tower, respectively. The surveillance station must ensure vehicular and driver control and safety. Saving data on the key system of the vehicle along with the monitoring of driving potential are essential in cases of emergency. Data regarding location and instrumental indicators must be collected to analyse what happened prior to the event. Without forfeiting functionality, the server-side logic handlers are permitted to scale to nodes by the application servers that rarely utilise synchronisation [100]. The work is transferred from these servers to the data processing server via two way communication and the database layer unloads its fragments through a straightforward interface with the servers. An embedded system integrated with an OBD module, RTC and GPS was prototyped to transfer real-time data to the server [103]. A microSD card was used to store data and pass it to the database for processing through data visualisation and extensive analysis codes. Navigation is executed through the quad-band GSM/GPRS communication module while the OBD diagnoses the vehicular system for its management and improvement.

The governing of traffic signal duration and the management of EVs are necessities of a well-functioning traffic management system. Vehicle supervision in deadlock situations is challenging due to which a system based on Wireless Sensor Networks (WSN) was implemented [106]. GPS and messaging services are employed to update the patient's condition in the EV to the central database. Computation of traffic volume and vehicle ID are performed using the WSNs and forwarded to all the access points en route to the hospital. The time span of green light is allocated based on the worst or best case scenarios and the duration is minimal when there is frequent movement of EVs to prevent starvation scenarios. Automated driving mechanisms-based control tactics focus on individual vehicles for safety purposes and help avoid congestion. RSUs are the signal controllers that retrieve wireless transmissions and employ lane switching algorithms using an Enhanced Intelligent Driver Module (EIDM) for vehicular control while ensuring avoidance of any endangering movement by the vehicle [111]. After data acquisition from the vehicles, the green light priority is determined by a pre-emption algorithm that helps reduce delay and enhance safety measures. The improvement of traffic and vehicular management can be obtained using an intelligent traffic congestion control (ITCC) approach that utilises fuzzy logic for preference [112]. The data packets are transferred with reduced wait duration through a congestion-aware routing algorithm (CARA). The RSU accumulates data from the WSN sensor nodes and passes it on to the Dynamic Traffic Management Center (DTMC) which initiates the fuzzy logic to decide the priority and allocate the required green signal duration. This approach provides facile deployment, inexpensive, lucid functionality and easy maintenance along with the optimisation of travel duration. A "Green Corridor"-based approach was implemented to enhance the smart traffic control and governance system. It employed the Worldwide Interoperability for Microwave Access (WiMAX) as the base station as it imparts a wide area coverage [117] and RFID was used to receive vehicular data and notify the traffic light to turn the signal green when an EV approaches until it reaches the destination. An IoT and multisensory data fusion-based architecture was implemented to guide EVs in avoiding congested routes and road obstructions [121]. The real time data of the EV is acquired using sensors which then transmits it to the microcontroller unit (MCU) enabled with Wi-Fi for the computation of speed and emissions of the vehicle to evaluate the intensity of the traffic via fuzzy logic. The severity of the traffic jam and the duration data are crowd sourced using an ARM application and the best possible route is acquired for smooth movement of the EV.

7.5 Proposed Methodology

Increase in the amount of vehicles on roads and decreased flexibility of free flow traffic are causing major life threats in the present world, leading to the loss of many lives every day. Most of the accidents caused on roads end up without proper medical reach. As the world is leading towards technological advancements, it tends to rely on automation and IoT that are embedded in all the devices used in our daily life. Thus, our objective is to introduce a smart board combined with modern IoT for the purpose of an emergency response system, placed in emergency vehicles and hospitals as depicted

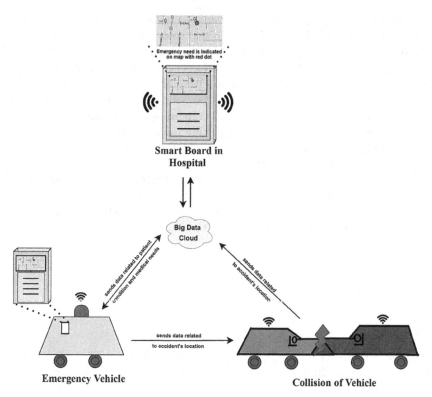

FIGURE 7.12
Proposed method.

in Figure 7.12. This smart IoT board keeps a track of the nearest emergency medical need and provides aid, both in terms of patient and vehicle health monitoring. The smart board placed in the hospitals will have track of the nearest emergency needs (i.e. indicators) and it also notifies the respective doctors to arrange for the required medical treatment. The board placed in the emergency vehicle will act as a source of communication between the traffic, the nearest hospital and also the vehicle that has been in the accident. The latest electric vehicles introduced in the automobile market come with the IoT-enabled technology where if the car meets with an accident, the airbags tend to blow up for protection. The trigger of the airbag being opened can be used as an indicator and a notification can be sent to the nearest hospital and to the closest emergency vehicle (connected to IoT). These hospitals and emergency vehicles can have the IoT smart board that helps in indicating the need for emergency medical service along with the proper location of the incident. The emergency vehicle can also be integrated with additional features like patient monitoring system and vehicle health monitoring system. Patient monitoring system is employed to monitor the patient's health in real-time and track the nearest hospital based on the location. It also supports in these recent times of pandemic due to Covid-19 where the shortage of oxygen cylinders is a major issue. Hence, the continuous monitoring of patient's health (integrated with IoT) can succour to find the nearest hospital having the required amount of oxygen, depending on the condition of the patient.

Vehicle Collision is a non-deterministic event. To detect this event various parameters and sensor data from the moving vehicle needs to be monitored continuously. The recorded data needs to be stored in a remote server so that the smart boards can use this data during the time of collision. At any instance of time, if the server goes down or if it fails to send the data to smart board, then smart boards will still be in an assumption that there is no occurrence of collision at this point of time which is risky for the patient. The processing of such huge data within seconds using conventional consumes a lot of time and computational power. These two problems can be solved using Hadoop. The data from sensors connected to the vehicle are transmitted and stored in a HDFS (Hadoop Distributed File System). Since HDFS stores the data on multiple nodes, even if a node shuts down the data can be retrieved using a different node. MapReduce helps the problem of processing this big data in less time interval. The data from HDFS can then be accessed by the remote Smart Boards in Hospitals and Emergency vehicles to find the closest path to the Hospital and the vehicle collided location and it consumes very less time and very productive in such real time applications.

7.6 Conclusion

In this chapter, we have presented the new trends in Emergency vehicle monitoring, health monitoring and patient monitoring. The need of IoT in tackling several problems for emergency vehicle is discussed. According to the various literature available, techniques are developed to overcome many problems related to the emergency response systems. But there is a necessity of giving utmost importance in technology advancements for automating and embedding IoT in the systems which provides reliable and secure operation. In this regard, we have proposed an IoT technology that helps in providing excellent service for patient and vehicle health monitoring. An IoT-enabled smart board can be introduced so that the emergency response activities can be forwarded to the emergency vehicles and hospitals when any accident occurs, an alert and necessary needs can be displayed in smart board. This automates in finding the nearest hospitals and alerting the medical personal about the event, so fast and ad hoc decisions can be taken.

References

1. T. T. Thakur, A. Naik, S. Vatari, and M. Gogate, "Real time traffic management using Internet of Things," *2016 International Conference on Communication and Signal Processing (ICCSP)*, 2016, pp. 1950–1953, doi: 10.1109/ICCSP.2016.7754512.
2. P. Rizwan, K. Suresh, and M. R. Babu, "Real-time smart traffic management system for smart cities by using Internet of Things and big data," *2016 International Conference on Emerging Technological Trends (ICETT)*, 2016, pp. 1–7, doi: 10.1109/ICETT.2016.7873660.
3. A. Saikar, M. Parulekar, A. Badve, S. Thakkar, and A. Deshmukh, "TrafficIntel: Smart traffic management for smart cities," *2017 International Conference on Emerging Trends & Innovation in ICT (ICEI)*, 2017, pp. 46–50, doi: 10.1109/ETIICT.2017.7977008.

4. Ramakrishna Nallathiga, Envisioning a comprehensive transport strategy for Mumbai. *Indian Journal of Transport Management*, 30(2) (June 1, 2006): pp. 153–177. SSRN: https://ssrn.com/abstract=987475.

5. Shashank Bharadwaj, Sudheer Ballare, and Munish K. Chandel, Impact of congestion on greenhouse gas emissions for road transport in Mumbai metropolitan region. *Transportation Research Procedia*, 25 (2017): 3538–3551. ISSN 2352-1465, doi: 10.1016/j.trpro.2017.05.282.

6. L. Li, S. Shrestha, and G. Hu, "Analysis of road traffic fatal accidents using data mining techniques," *2017 IEEE 15th International Conference on Software Engineering Research, Management and Applications (SERA)*, 2017, pp. 363–370, doi: 10.1109/SERA.2017.7965753.

7. Y. A. Seliverstov, S. A. Seliverstov, V. I. Komashinskiy, A. A. Tarantsev, N. V. Shatalova, and V. A. Grigoriev, "Intelligent systems preventing road traffic accidents in megalopolises in order to evaluate," *2017 XX IEEE International Conference on Soft Computing and Measurements (SCM)*, 2017, pp. 489–492, doi: 10.1109/SCM.2017.7970626.

8. S. Kannan, G. Dhiman, Y. Natarajan, A. Sharma, S. N. Mohanty, M. Soni, U. Easwaran, H. Ghorbani, A. Asheralieva, and M. Gheisari, Ubiquitous vehicular Ad-Hoc network computing using deep neural network with IoT-based bat agents for traffic management. *Electronics*, 10 (2021): 785, doi: 10.3390/electronics10070785.

9. P. Soleimani, M. R. B. Marvasti, and P. Ghorbanzadeh, "A hybrid traffic management method based on combination of IOV and VANET network in urban routing for emergency vehicles," *2020 4th International Conference on Smart City, Internet of Things and Applications (SCIOT)*, 2020, doi: 10.1109/sciot50840.2020.92501.

10. N. Stanton (Ed.). Advances in human aspects of transportation. *Advances in Intelligent Systems and Computing* (2020), doi: 10.1007/978-3-030-50943-9.

11. F. Li et al., "Different traffic density connectivity probability analysis in VANETs with measured data at 5.9 GHz," *Proceedings of 2018 16th International Conference on Intelligent Transport System Telecommunications ITST*, 2018, pp. 1–7.

12. A. Mohanty, S. Mahapatra, and U. Bhanja, Traffic congestion detection in a city using clustering techniques in VANETs. *Indonesian Journal of Electrical Engineering and Computer Science*, 13(3) (2019): 884–891.

13. H. Sitohang, and Merni, S. K-means method for analysis of accident-prone areas in Palangka Raya. *Journal of Physics: Conference Series* 1566 (2020): 012008.

14. S. Askari, Fuzzy C-means clustering algorithm for data with unequal cluster sizes and contaminated with noise and outliers: Review and development. *Expert Systems with Applications* (2020): 113856, doi: 10.1016/j.eswa.2020.113856.

15. F. Nie, X. Zhao, R. Wang, X. Li, and Z. Li, Fuzzy K-means clustering with discriminative embedding. *IEEE Transactions on Knowledge and Data Engineering* (2020): 1–1, doi: 10.1109/tkde.2020.2995748.

16. M. S. B. Syed, F. Memon, S. Memon, and R. A. Khan, "IoT based emergency vehicle communication system," *2020 International Conference on Information Science and Communication Technology (ICISCT)*, 2020, pp. 1–5, doi: 10.1109/ICISCT49550.2020.9079940.

17. F. Bhatti, M. A. Shah, C. Maple, and S. U. Islam, A novel Internet of Things-enabled accident detection and reporting system for smart city environments. *Sensors*, 19(9) (2019): 2071, doi: 10.3390/s19092071.

18. L. Broccardo, F. Culasso, and S. G. Mauro, Smart city governance: Exploring the institutional work of multiple actors towards collaboration, *International Journal of Public Sector Management*, 32(4) (2019): 367e387, doi: 10.1108/IJPSM-05-2018-0126.

19. A. Camero and E. Alba, Smart city and information technology: A review. *Cities* 93(May) (2019): 84e94, doi: 10.1016/j.cities.2019.04.014.

20. M. Gohar, M. Muzammal, and A. Ur Rahman, Smart TSS: Defining transportation system behavior using big data analytics in smart cities. *Sustainable Cities and Society*, 41(August 2017) (2018): 114e119, doi: 10.1016/j.scs.2018.05.008.

21. W. H. Lee and C. Y. Chiu, Design and implementation of a smart traffic signal control system for smart city applications. *Sensors*, 20(2) (2020), doi: 10.3390/s20020508.

22. Z. A. Almusaylim and N. Z. Jhanjhi, Comprehensive review: Privacy protection of user in location-aware services of mobile cloud computing. *Wireless Personal Communications*, 111(1) (2020): 541e564.

23. R. K. Yadav, R. Jain, S. Yadav, and S. Bansal, "Dynamic traffic management system using neural network based IoT system," *2020 4th International Conference on Intelligent Computing and Control Systems (ICICCS)*, 2020, doi: 10.1109/iciccs48265.2020.9121.

24. Y. Yang, J. Cui, and X. Cui, Design and analysis of magnetic coils for optimizing the coupling coefficient in an electric vehicle wireless power transfer system. *Energies*, 13(16) (2020): 4143, doi: 10.3390/en13164143.

25. H. Li, H. Dong, L. Jia, D. Xu, and Y. Qin, "Some practical vehicle speed estimation methods by a single traffic magnetic sensor," *2011 14th International IEEE Conference on Intelligent Transportation Systems (ITSC)*, October, 2011, IEEE, pp. 1566–1573, doi: 10.1109/ITSC.2011.6083076.

26. K. Nellore and G. Hancke, A survey on urban traffic management system using wireless sensor networks. *Sensors*, 16(2) (2016): 157, doi: 10.3390/s16020157.

27. M. Sarrab, S. Pulparambil, and M. Awadalla, Development of an IoT based real-time traffic monitoring system for city governance. *Global Transitions*, 2 (2020): 230–245.

28. Alternate route handbook. Federal Highway Administration, U.S. Dept. of Transportation, 2006, https://ops.fhwa.dot.gov/publications/ar_handbook/index.htm.

29. D. L. Hall and J. Llinas, An introduction to multisensor data fusion. *Proc IEEE*, 85(1) (1997): 6–23, doi: 10.1109/5.554205.

30. Vittorio Astarita, Vincenzo Pasquale Giofrè, Giuseppe Guido, and Alessandro Vitale, A review of traffic signal control methods and experiments based on Floating Car Data (FCD). *Procedia Computer Science*, 175 (2020): 745–751.

31. J. E. Naranjo, F. Jimenez, F. J. Serradilla, and J. G. Zato, Floating car data augmentation based on infrastructure sensors and neural networks. *IEEE Transactions on Intelligent Transportation Systems*, 13(1) (2012): 107–114, doi: 10.1109/TITS.2011.2180377.

32. D. Luxen and C. Vetter, "Real-time routing with OpenStreetMap data," *Proceedings of the 19th ACM SIGSPATIAL International Conference on Advances in Geographic Information Systems*, New York, NY, USA, ACM, 2011, pp. 513–516, doi: 10.1145/2093973.2094062.

33. Rout, R. R., Vemireddy, S., Raul, S. K., and Somayajulu, D. V. L. N. (2020). Fuzzy logic-based emergency vehicle routing: An IoT system development for smart city applications. *Computers & Electrical Engineering*, 88, 106839, doi: 10.1016/j.compeleceng.2020.10.

34. Mustafa Maad Hamdi, Lukman Audah, Sami Abduljabbar Rashid, and Mahmoud Al Shareeda, Techniques of early incident detection and traffic monitoring centre in VANETs: A review. *Journal of Communication*, 15(12) (2020): 896–904.

35. S. Manipriya, C. Mala, and Samson Mathew, A collaborative framework for traffic information in vehicular Adhoc network applications. *Journal of Internet Services and Information Security (JISIS)*, 10(3) (2020): 93–109.

36. J. Guerrero-Ibáñez, S. Zeadally, and J. Contreras-Castillo, Sensor technologies for intelligent transportation systems. *Sensors*, 18(4) (2018): 1212, doi: 10.3390/s18041212.

37. M. Mala, A. Sathya, S. Prathipa, and V. Kanimozhi, IOT Based Smart Integrated Medical Health Monitoring System, *ECLECTIC - 2020 Conference Proceedings, International Journal of Engineering Research & Technology (IJERT)*.

38. Ana Lavalle, Miguel A. Teruel, Alejandro Maté, and Juan Trujillo, Improving sustainability of smart cities through visualization techniques for big data from IoT devices. *Sustainability*, 12(14) (2020): 5595, doi: 10.3390/su12145595.

39. G. Karmakar, A. Chowdhury, J. Kamruzzaman, and I. Gondal, A smart priority-based traffic control system for emergency vehicles. *IEEE Sensors Journal*, 21(14) (2021): pp. 15849–15858, doi: 10.1109/JSEN.2020.3023149.

40. Andrey Volkov, IoT traffic prediction with neural networks learning based on SDN infrastructure. In *Distributed Computer and Communication Networks* (pp. 64–76). Springer, Cham. Moscow, Russia, 2020, September.

41. S. Majumdar, M. M. Subhani, B. Roullier, A. Anjum, and R. Zhu, Congestion prediction for smart sustainable cities using IoT and machine learning approaches. *Sustainable Cities and Society* 64 (2020): 102500, doi: 10.1016/j.scs.2020.102500.

42. Andrey Abdellah, Deep learning with long short-term memory for IoT traffic prediction. In *Internet of Things, Smart Spaces, and Next Generation Networks and Systems* (pp. 267–280). Springer, Cham, 2020.

43. Valeriy Elkin, IoT in traffic management: Review of existing methods of road traffic regulation. In *Applied Informatics and Cybernetics in Intelligent Systems* (pp. 536–551). Springer International Publishing, 2020.

44. L. Rui, Y. Zhang, H. Huang, and X. Qiu, A new traffic congestion detection and quantification method based on comprehensive fuzzy assessment in VANET. *KSII Transactions on Internet & Information Systems*, 12(1), 2018: 41–60.

45. J. Jung, S. Chun, X. Jin, and K.-H. Lee, Quantitative computation of social strength in social Internet of Things. *IEEE Internet of Things Journal*, 5(5) (2018): 4066–4075.

46. V. L. Tran, A. Islam, J. Kharel, and S. Y. Shin, "On the application of social Internet of Things with fog computing: A new paradigm for traffic information sharing system," *2018 IEEE 6th International Conference on Future Internet of Things and Cloud (FiCloud)*, 2018, pp. 349–354. IEEE.

47. Z. Lin and L. Dong, Clarifying trust in social Internet of Things. *IEEE Transactions on Knowledge and Data Engineering*, 30(2) (2018): pp. 234–248.

48. O. Younis and N. Moayeri, Employing cyber-physical systems: Dynamic traffic light control at road intersections. *IEEE Internet of Things Journal*, 4(6) (2017): 2286–2296.

49. L. Atzori, A. Floris, R. Girau, M. Nitti, and G. Pau, Towards the implementation of the social internet of vehicles. *Computer Networks*, 147 (2018): 132–145.

50. K. Zia, A. Muhammad, A. Khalid, A. Din, and A. Ferscha, Towards exploration of social in social internet of vehicles using an agent-based simulation. *Complexity*, 2019 (2019).

51. C. Menelaou, S. Timotheou, P. Kolios, C. G. Panayiotou, and M. M. Polycarpou, Minimizing traffic congestion through continuous-time route reservations with travel time predictions. *IEEE Transactions on Intelligent Vehicles*, 4(1) (2019): 141–153.

52. M. S. Roopa, Ayesha Siddiq, Rajkumar Buyya, K. R. Venugopal, S. S. Iyengar, and L. M. Patnaik. Dynamic management of traffic signals through social IoT. *Procedia Computer Science*, 171 (2020): 1908–1916.

53. Y. Wang, H. Wang, J. Xuan, and D. Y. C. Leung, Powering future body sensor network systems: A review of power sources. *Biosensors and Bioelectronics*, 166, (2020): 112410, doi: 10.1016/j.bios.2020.112410.

54. R. Sokullu, M. Akkaş, and H. Cetin, "Wireless patient monitoring system", *Fourth International Conference on Sensor Technologies and Applications*, 2010, pp. 179–184.

55. J. Jung, S. Shin, M. Kang, K. H. Kang, and Y. T. Kim, Development of wearable wireless electrocardiogram detection system using Bluetooth low energy. *Electronics*, 10 (2021): 608, doi: 10.3390/electronics10050608.

56. Nižetić, Sandro, Petar Šolić, Diego López-de-Ipiña González-De, and Luigi Patrono, Internet of Things (IoT): Opportunities, issues and challenges towards a smart and sustainable future. *Journal of Cleaner Production*, 274 (2020): 122877, doi: 10.1016/j.jclepro.2020.122877.

57. E. K. Svavarsdottir, S. W. Kamban, E. Konradsdottir, and A. O. Sigurdardottir, The impact of family strengths oriented therapeutic conversations on parents of children with a New Chronic illness diagnosis. *Journal of Family Nursing*, doi: 10.1177/1074840720940674.

58. Prosanta Gope and Tzonelih Hwang, BSN-care: A secure IoT-based modern healthcare system using body sensor network. *IEEE Sensors Journal*, 16(5) (2015): 1368–1376.

59. J. W. P. Ng, B. P. L. Lo, O. Wells, M. Sloman, N. Peters, A. Darzi, C. Toumazou, and G.-Z. Yang, "Ubiquitous monitoring environment for wearable and implantable sensors (UbiMon)," *Proceedings of 6th International Conference on Ubiquitous Computing (UbiComp'04)*, Nottingham, UK, 7–14 September 2004.

60. D. Kreps, T. Komukai, T. V. Gopal, and K. Ishii (Eds.), "Human-centric computing in a data-driven society," *IFIP Advances in Information and Communication Technology* (2020), doi: 10.1007/978-3-030-62803-1.

61. K. Jose Reena and R. Parameswari, "IOT based health tracking shoe for elderly people using gait monitoring system," *2021 7th International Conference on Advanced Computing and Communication Systems (ICACCS)*, 2021, pp. 1701–1705, doi: 10.1109/ICACCS51430.2021.9441754.

62. S. Dagtas, G. Pekhteryev, and Z. Sahinoglu, Multistage real time health monitoring via ZigBee in smart homes. In *21st International Conference on Advanced Information Networking and Applications Workshops (AINAW'07)* (Vol. 2, pp. 782–786). IEEE, 2007.

63. A. Sabban, New compact wearable metamaterials circular patch antennas for IoT, medical and 5G applications. *Applied System Innovation*, 3(4) (2020): 42, doi: 10.3390/asi3040042.

64. M. Alper Akkaş, Radosveta Sokullu, and H. Ertürk Çetin, Healthcare and patient monitoring using IoT. *Internet of Things*, 11 (2020): 100173.

65. W. Huifeng, S. N. Kadry, and E. D. Raj, Continuous health monitoring of sportsperson using IoT devices based wearable technology. *Computer Communications* (2020), doi: 10.1016/j.comcom. 2020.04.025.

66. J. Shen, S. Chang, J. Shen, Q. Liu, and X. Sun, A lightweight multi-layer authentication protocol for wireless body area networks. *Future Generation Computer Systems*, 78 (2018): 956–963.

67. S. Dagtas, G. Pekhteryev, and Z. Sahinoglu, Multistage real time health monitoring via ZigBee in smart homes. Mitsubishi Electric Research Laboratories (2007).

68. M. Ghamari, B. Janko, R. Sherratt, W. Harwin, R. Piechockic, and C. Soltanpur, A survey on wireless body area networks for eHealthcare systems in residential environments. *Sensors*, 16(6) (2016): 831, doi: 10.3390/s1606083.

69. R. Y. Kim, The impact of COVID-19 on consumers: Preparing for digital sales. *IEEE Engineering Management Review*, 48(3) (2020): 212–218, doi: 10.1109/EMR.2020.2990115.

70. Singh Ravi Pratap, Mohd Javaid, and Abid Haleem, Internet of things (IoT) applications to fight against COVID-19 pandemic. *Diabetes & Metaboli Syndrome: Clinical Research & Reviews*, 14(4) (July–August 2020): 521–524.

71. Ravi Pratap Singh, Mohd Javaid, Abid Haleem, Raju Vaishya, and Shokat Ali, Internet of Medical Things (IoMT) for orthopaedic in COVID-19 pandemic: Roles, challenges, and applications. *Journal of Clinical Orthopaedics and Trauma*, in press, corrected proof Available online 15 May 2020. https://doi.org/10.1016%2Fj.jcot.2020.05.011.

72. P. Chatterjee, L. J. Cymberknop, and R. L. Armentano, "IoT-based decision support system for intelligent healthcare — applied to cardiovascular diseases," *2017 7th International Conference on Communication Systems and Network Technologies (CSNT)*, Nagpur, 2017, pp. 362–366, doi: 10.1109/CSNT.2017.8418567.

73. T. McDermott, "Developing systems thinking skills using healthcare as a case study," *2018 13th Annual Conference on System of Systems Engineering (SoSE)*, Paris, 2018, pp. 240–244, doi: 10.1109/SYSOSE.2018.8428717.

74. L. Garg, E. Chukwu, N. Nasser, C. Chakraborty, and G. Garg, Anonymity preserving IoT-based COVID-19 and other infectious disease contact tracing model. *IEEE Access* (2020): 1–1, doi: 10.1109/access.2020.3020513.

75. G. S. Aujla and A. Jindal, A decoupled blockchain approach for edge-envisioned IoT-based healthcare monitoring. *IEEE Journal on Selected Areas in Communications* (2020): 1–1, doi: 10.1109/jsac.2020.3020655.

76. M. Abdur Rahim, M. Arafatur Rahman, M. M. Rahman, A. Taufiq Asyhari, M. Zakirul Alam Bhuiyan, and D. Ramasamy, Evolution of IoT-enabled connectivity and applications in automotive industry: A review. *Vehicular Communications* (2020): 100285, doi: 10.1016/j.vehcom.2020.100285.

77. P. K. Binu, K. J. Sredhey, and R. S. Anuvind, "An Iot based safety and security mechanism for passenger vehicles," *2019 2nd International Conference on Intelligent Computing, Instrumentation and Control Technologies (ICICICT)*, 2019, pp. 1502–1506, doi: 10.1109/ICICICT46008.2019.899 3177.

78. Y. Huo, W. Tu, Z. Sheng, and V. C. M. Leung, "A survey of in-vehicle communications: Requirements, solutions and opportunities in IoT," *2015 IEEE 2nd World Forum on Internet of Things (WF-IoT)*, 2015, pp. 132–137, doi: 10.1109/WF-IoT.2015.7389040.

79. Xueli Feng and Jie Hu, Research on the identification and management of vehicle behaviour based on Internet of things technology. *Computer Communications*, 156 (2020): 68–76, doi: 10.1016/j.comcom.2020.03.035.

80. A. Srinivasan, "IoT cloud based real time automobile monitoring system," *2018 3rd IEEE International Conference on Intelligent Transportation Engineering (ICITE)*, 2018, pp. 231–235, doi: 10.1109/ICITE.2018.8492706.

81. Gheorghe Panga, Sorin Zamfir, Titus Bălan, and Ovidiu Popa, Iot diagnostics for connected cars. Transilvania University, Braşov, Romania, doi: 10.19062/2247-3173.2016.18.1.39.

82. Nishita Goyal, Vibhu Goel, Manan Anand, and Sahil Garg, Smart vehicle: Online prognosis for vehicle health monitoring. *Department of Electronics and Communication Engineering*, 9(2) (January–June 2020). ISSN: 2278-0947.

83. A. F. A. Hamid, M. T. A. Rahman, S. F. Khan, A. H. Adom, M. A. Rahim, N. A. Rahim, M. H. N. Ismail, and A. Norizan, Connected car: Engines diagnostic via Internet of Things (IoT). *Journal of Physics: Conference Series*, 908(1) (2017): 012079, doi: 10.1088/1742-6596/908/1/012079.

84. F. Fahmi, F. Nurmayadi, B. Siregar, M. Yazid, and E. Susanto, Design of hardware module for the vehicle condition monitoring system based on the Internet of Things. *IOP Conference Series: Materials Science and Engineering*, 648 (2019): 012039, doi: 10.1088/1757-899X/648/1/012039.

85. B. V. P. Prasad, J. Tang, and S. Luo, "Design and implementation of SAE J1939 vehicle diagnostics system," *2019 IEEE International Conference on Computation, Communication and Engineering (ICCCE)*, 2019, pp. 71–74, doi: 10.1109/ICCCE48422.2019.9010769.

86. Ercument Turk, and Moharram Challenger, "An android-based IoT system for vehicle monitoring and diagnostic," *[IEEE 2018 26th Signal Processing and Communications Applications Conference (SIU) - Izmir, Turkey (2018.5.2–2018.5.5)] 2018 26th Signal Processing and Communications Applications Conference (SIU)*, 2018, pp. 1–4, doi: 10.1109/SIU.2018.8404378.

87. Xuan Shao, Xingwu Kang, Xuping Wang, and Xiaojing Yuan, Design of special vehicle condition monitoring system based on J1939. *Journal of Physics: Conference Series*, 1549(3) (2020): 032092, doi: 10.1088/1742-6596/1549/3/032092.

88. Sudam Pawar, Shubham Admuthe, Mugdha Shah, Sachita Kulkarni, and Rahul Thengadi, Pothole Detection and Warning System using IBM Bluemix Platform. e-ISSN: 2395-0056, 03(11) (November 2016): p-ISSN: 2395-0072.

89. E. Husni, G. B. Hertantyo, D. W. Wicaksono, F. C. Hasibuan, A. U. Rahayu, and M. A. Triawan, "Applied Internet of Things (IoT): Car monitoring system using IBM BlueMix," *2016 International Seminar on Intelligent Technology and Its Applications (ISITIA)*, 2016, pp. 417–422, doi: 10.1109/ISITIA.2016.7828696.

90. Gurdit Singh, Divya Bansal, and Sanjeev Sofat, A smartphone-based technique to monitor driving behavior using DTW and crowdsensing. *Pervasive and Mobile Computing*, 40 (2017): 56–70. S1574119216301250, doi: 10.1016/j.pmcj.2017.06.003.

91. Muhammad Farrukh Yaqub and Iqbal Gondal, "Smart phone-based vehicle condition monitoring," *[IEEE 2013 IEEE 8th Conference on Industrial Electronics and Applications (ICIEA 2013) - Melbourne, VIC (2013.6.19–2013.6.21)] 2013 IEEE 8th Conference on Industrial Electronics and Applications (ICIEA)*, 2013, pp. 267–271, doi: 10.1109/ICIEA.2013.6566378.

92. P. Varaiya and S. Shladover, "Sketch of an IVHS system architecture," *Vehicle Navigation and Information System, IEEE Conference 91*, Ann Arbor, July, 1992.

93. S. Shladover, "Potential freeway capacity effects of automated vehicle control systems," *Applications of Advanced Technologies in Transportation Engineering, ASCE Conference*, Minneapolis, MN, August 1991, pp. 213–217.

94. W. Zhang, "Vehicle health monitoring for AVCS malfunction management," *Proceedings of VNIS '93 - Vehicle Navigation and Information Systems Conference*, 1993, pp. 501–504, doi: 10.1109/VNIS.1993.585681.

95. O. Benedettini, T. S. Baines, H. W. Lightfoot, and R. M. Greenough, State-of-the-art in integrated vehicle health management. *Proceedings of the Institution of Mechanical Engineers, Part G: Journal of Aerospace Engineering,* 223(2) (2009): 157–170, doi: 10.1243/09544100JAERO446.

96. Z. Williams, "Benefits of IVHM: An analytical approach," *In Proceedings of the 2006 IEEE Aerospace Conference, Big Sky,* Montana, USA, 4–11 March 2006, paper no. 1507.

97. K. M. Janasak and R. R. Beshears, "Diagnostics to prognostics – a product availability technology evolution," *Proceedings of the 2007 Reliability and Maintainability Symposium – RAMS'07,* Orlando, Florida, USA, 22–25 January 2007, pp. 113–118.

98. E. Baroth, W. T. Powers, J. Fox, B. Prosser, J. Pallix, K. Schweikard, and, J. Zakrajsek, "IVHM (Integrated Vehicle Health Management) techniques for future space vehicles," *Proceedings of the 37th AIAA/ASME/SAE/ASEE Joint Propulsion Conference Exhibit, Salt Lake City,* Utah, USA, 8–11 July 2001, AIAA paper 2001-3523.

99. H. Lingling, L. Haifeng, X. Xu, and L. Jian, "An intelligent vehicle monitoring system based on Internet of Things," *2011 Seventh International Conference on Computational Intelligence and Security,* 2011, pp. 231–233, doi: 10.1109/CIS.2011.59.

100. A. S. Chesnokov, M. G. Gorodnichev, K. A. Gavrish, and M. A. Zhidkova, "Intelligent vehicle condition monitoring system," *2019 Systems of Signals Generating and Processing in the Field of on Board Communications,* 2019, pp. 1–4, doi: 10.1109/SOSG.2019.8706727.

101. M. G. Gorodnichev and A. N. Nigmatulin, Technical and program aspects on monitoring of highway flows (case study of Moscow city). *Advances in Intelligent Systems and Computing,* 224 (2013): 195–204.

102. A. Srinivasan, "IoT cloud based real time automobile monitoring system," *2018 3rd IEEE International Conference on Intelligent Transportation Engineering (ICITE),* 2018, pp. 231–235, doi: 10.1109/ICITE.2018.8492706.

103. Alexandros Mouzakitis, Nataraja Muniyappa, Richard Parker, and Shamal Puthiyapurayil, "Advanced automated on board vehicle diagnostics testing," *UKACC International Conference on Control,* 2010.

104. Shi-Huang Chen, Jhing-Fa Wang, YuRu Wei, John Shang, and Shao-Yu Kao, "The implementation of real-time online vehicle diagnostics and early fault estimation system", *2011 Fifth International Conference on Genetic and Evolutionary Computing,* 2011, Kinmen, Taiwan.

105. J. A. M. Polar, D. S. Silva, A. L. Fortunato, L. A. C. Almeida, and C. A. Dos Reis Filho, "Bluetooth sensor network for remote diagnostics in vehicles", *2003 IEEE International Symposium on Industrial Electronics (Cat. No. 03TH8692),* 2003, Rio de Janeiro, Brazil.

106. W. Kang, G. Xiong, L. Yisheng, X. Dong, F. Zhu, and Q. Kong, "Traffic signal coordination for emergency vehicles," *17th International IEEE Conference on Intelligent Transportation Systems (ITSC),* 2014, pp. 157–161, doi: 10.1109/ITSC.2014.6957683.

107. J. Feng and H. Zhao, Dynamic nodes collaboration for target tracking in wireless sensor networks. *IEEE Sensors Journal,* doi: 10.1109/JSEN.2021.3093473.

108. C. S. Partha, S. Arti, and S. Raj, "Adaptive and optimized emergency vehicle dispatching algorithm for intelligent traffic management system," *Third International Conference on Recent Trends in Computing,* 2015, vol. 57, doi: 10.1016/j.procs.2015.07.454.

109. J. So, J. Kang, S. Park, I. Park, and J. Lee, Automated emergency vehicle control strategy based on automated driving controls. *Hindawi's Journal of Advanced Transportation,* 2020 (2020), doi: 10.1155/2020/3867921.

110. J. Barrachina, P. Garrido, M. Piedad, F. Martinez, J. Cano, C. Calafate, and P. Manzoni, Road side unit deployment: A density-based approach. *IEEE Intelligent Transportation Systems Magazine* (2013), doi: 10.1109/MITS.2013.2253159.

111. A. Kesting, M. Treiber, and D. Helbing, Enhanced intelligent driver module to access the impact of driving strategies on traffic capacity. *Series A, Mathematical, Physical and Engineering Sciences* (2010), doi: 10.1098/rsta.2010.0084.

112. M. Shelke, A. Malhotra, and P. N. Mahalle, Fuzzy priority based intelligent traffic congestion control and emergency vehicle management using congestion-aware routing algorithm. *Journal of Ambient Intelligence and Humanized Computing* (2019), doi: 10.1007/s12652-019-01523-8.

113. L. P. J. Rani, M. K. Kumar, K. S. Naresh, and S. Vignesh, "Dynamic traffic management system using infrared (IR) and Internet of Things (IoT)," *2017 Third International Conference on Science Technology Engineering & Management (ICONSTEM)*, 2017, pp. 353–357, doi: 10.1109/ICONSTEM. 2017.8261308.

114. A. Mittal and D. Bhandari, "A novel approach to implement green wave system and detection of stolen vehicles," *IEEE Sensors Journal* 15, no. 2 (2014): 1109–1113.

115. R. Sundar, S. Hebbar, and V. Golla, Implementing intelligent traffic control system for congestion control, ambulance clearance and stolen vehicle detection. *IEEE Sensors Journal*, 15(2) (2014): 1109–1113.

116. B. Rajak and D. S. Kushwaha, Traffic control and management over IoT for clearance of emergency vehicle in smart cities. In Fong, S., Akashe, S., Mahalle, P. (eds.), *Information and Communication Technology for Competitive Strategies*. Lecture Notes in Networks and Systems, vol. 40, Springer, Singapore, 2019, doi: 10.1007/978-981-13-0586-3_12.

117. A. I. Hammoodi, A. S. Kashkool, and H. Raad, "Flexible dual-band slot antenna for WLAN and WiMAX applications," *2021 IEEE 19th International Symposium on Antenna Technology and Applied Electromagnetics (ANTEM)*, 2021, pp. 1–2, doi: 10.1109/ANTEM51107.2021.9518919.

118. M. El-Hadid and Y. E. S. B. Yaseer, "Realistic chipless RFID tag modeling, mathematical framework and 3D EM simulation," *2019 IEEE International Conference on RFID Technology and Applications (RFID-TA)*, 2019, pp. 201–206, doi: 10.1109/RFID-TA.2019.8892178.

119. B. Rashid and M. H. Rehmani, Applications of wireless sensor networks for urban area: A survey. *Journal of Network and Computer Applications*, 60 (2016): 192–219, doi: 10.1016/j.jnca.2015. 09.008.

120. D. L. Hall and J. Llinas, An introduction to multisensor data fusion. *Proceedings of the IEEE* (1997): 06–23, doi: 10.1109/5.554205.

121. J. E. Naranjo, F. Jimenez, F. J. Serradilla, and J. G. Zato, Floating car data augmentation based on infrastructure sensors and neural networks. *IEEE Transactions on Intelligent Transportation Systems*, 13 (2012): 107–114, doi: 10.1109/TITS.2011.2180377.

122. A. S. Handayani, H. M. Putri, S. Soim, N. L. Husni, R. Rusmiasih, and C. R. Sitompul, "Intelligent transportation system for traffic accident monitoring," *2019 International Conference on Electrical Engineering and Computer Science*, 2019, pp. 156–161, doi: 10.1109/ICECOS47637.2019.8984525.

123. A. Shamshad and I. U. Haq, "A parallelized data processing algorithm for map matching on Open Source Routing Machine (Osrm) server", *2020 14th International Conference on Open Source Systems and Technologies (ICOSST)*, 2020, pp. 1–6, doi: 10.1109/ICOSST51357.2020.9333085.

124. Z. Ahmed, A. Rawat, and P. Kumari, An analysis of IoT based smart cities. *International Journal of Engineering Trends and Applications (IJETA)*, 8(4) (2021).

125. Y. S. Jeong and J. H. Park, IoT and smart city technology: Challenges, opportunities, and solutions. *Journal of Information Processing Systems*, 15 (2019): 233–238.

126. R. J. Hassan, S. R. M. Zeebaree, S. Y. Ameen, S. F. Kak, M. A. M. Sadeeq, Z. S. Ageed, A. AL-Zebari, and A. A. Salih, State of art survey for IoT effects on smart city technology: Challenges, opportunities, and solutions. *Asian Journal of Research in Computer Science*, 8(3) (2021): 32–48, doi: 10.9734/ajrcos/2021/v8i330202.

127. S. K. Rao and R. Prasad, Impact of 5G technologies on smart city implementation. *Wireless Personal Communications*, 100 (2018): 161–176, doi: 10.1007/s11277-018-5618-4.

128. Susan Rachmawati, Arman Syah Putra, Abednego Priyatama, Dudi Parulian, Dona Katarina, Muhammad Tri Habibie, Matdio Siahaan, Endah Prawesti Ningrum, Alsen Medikano, and V. H. Valentino, "Application of drone technology for mapping and monitoring of corn agricultural land," *2021 International Conference on ICT for Smart Society (ICISS)*, 2021, pp. 1–5. IEEE.

129. Martin Bauer, Luis Sanchez, and JaeSeung Song, IoT-enabled smart cities: Evolution and outlook. *Sensors*, 21(13) (2021): 4511.

130. S. P. Mohanty, U. Choppali, and E. Kougianos, Everything you wanted to know about smart cities: The internet of things is the backbone. *IEEE Consumer Electronics Magazine*, 5(3) (2016): 60–70.

131. A. B. Sallow, H. I. Dino, Z. S. Ageed, M. R. Mahmood, and M. B. Abdulrazaq, Client/Server remote control administration system: Design and implementation. *International Journal of Multidisciplinary Research Publishing*, 3 (2020): 7.
132. Zainab Salih Ageed, Subhi RM Zeebaree, Mohammed Mohammed Sadeeq, Shakir Fattah Kak, Zryan Najat Rashid, Azar Abid Salih, and Wafaa M. Abdullah, A survey of data mining implementation in smart city applications. *Qubahan Academic Journal*, 1 (2021): 91–99, doi: 10.48161/qaj.v1n2a52.

8

Pandemic Management Using Internet of Things and Big Data – A Security and Privacy Perspective

K. S. Arvind
Jain University, Jakkasandra, India

S. Vanitha
PES University, Bangalore, India

K. S. Suganya
Bannari Amman Institute of Technology, Sathyamangalam, India

CONTENTS

8.1 Introduction 159
8.2 Related Work 160
8.3 IoT in Pandemic Management 163
    8.3.1 Role of Big Data in Pandemic Management 164
    8.3.2 Role of Contact Tracing Applications in Pandemic Management 164
    8.3.3 Role of Other IoT Devices in Pandemic Management 165
    8.3.4 Role of IoT Empowered Smart Hospitals 166
8.4 Security and Privacy Requirements of IoT in Pandemic Management 168
8.5 Conclusion and Future Focus 170
References 170

8.1 Introduction

In COVID-19, contact tracing involves identifying individuals who have been in contact with affected individual. This process of contact tracing prevents the community spreading of the disease. Many governments are trying to identify the infected individuals using contact tracing mobile applications that are equipped with GPS, Bluetooth or NFC devices. The Chinese government has encouraged its citizens to use a mobile application named "HealthCode" for contact tracing. The application generally displays a color code (Green/ Yellow/Red code for unrestricted/7 days' quarantine/14 days' quarantine) with the data from user's travel history and health status using Big Data analytics. The Singaporean government has introduced "Tracetogether" application which is a Bluetooth enable Mobile application. This application works by sharing identity-anonymous messages between users that have active application in their mobile phones to detect the infected individual or those who have been in contact with infected individual or those who have been enforced quarantine. The South Korean government has launched an application "Co100" which alerts the Korean citizens using application about infected individual when they are in proximity of 100 meters with the infected individual. Indian government have also launched an application named "Aarogya Setu" and has been made mandatory for the government and private sector employees and works similar to Tracetogether application.

While the above methods employed various technologies to advance contact tracing in these pandemic times, the reuse of massive infrastructure of Internet of Things that has been already employed in smart cities can improve the contact tracing to its next level. Such Infrastructure contains powerful devices that have led to the improvement of contact tracing mobile applications.

Machine learning, data mining, Big Data and deep learning are all examples of artificial intelligence (AI). Data mining techniques are currently being used to detect and predict a wide range of tasks, including disease prediction. Disease outbreaks could be tracked in real time using Big Data. COVID-19 is different from earlier pandemics in terms of the number of new infections in the country. Big Data is a notion that takes a big amount of data and gives it new meaning and value. For many processes, the appraisal and utilization of sufficient data are crucial. For information retrieval and extraction, Big Data is available in both organized and unstructured formats. Big Data has recently grown significantly, with exponential growth due to data gathered through machine learning and used for applications such as sales analytics, stock market prediction Big Data, food reviews through sentiment analysis, cloud computing, movie recommendation systems, deep learning regarding leukemia diseases, fake profiles, flight web search analytics, Cricket match winning prediction and IoT threads for predicting Denial of Service attacks using IoT threads The volume, speed, diversity, value and integrity of information known are all characteristics of Big Data. Complex human pandemic-related techniques and reactions can be aided by digital health technology.

In this chapter, we are going to address the necessity of IoT in pandemic management and ensuring that security and privacy is employed in these Covid-tracing applications based on the massive infrastructure of smart cities existing Internet of Things devices and technologies. The following discussion is organized as follows. Section 8.2 discusses study and techniques employed for security and privacy in IoT and contact tracing. Section 8.3 discusses in detail about the role of contact tracing applications, employing smart hospitals and employing various IoT-related smart solutions for pandemic management. Section 8.4 discusses the various aspects and requirements for ensuring security and privacy while employing Internet of Things for pandemic management. Section 8.5 concludes this study.

8.2 Related Work

While ensuring security and privacy for Internet of Things, many symmetric and asymmetric cryptographic techniques have been employed since it advent. We will be discussing some of the most related works in relation with security of IoT and contact tracing application. With considering collaborative real-world data sharing scenario, Fugkeaw and Sato [1] have implemented secure access control model. It is an integration model with components from "Role-based Access Control Model (RBAC)" and "Cipher Text Policy-Attribute-based Encryption (CP-ABE)". The method proved a better collusion resistance and data sharing for big files.

Few others have enhanced the existing CP-ABE models with many levels of authority, and they appear to be a viable way for restricting access to large amounts of Big Data stored in the cloud [2]. This implementation eliminated the need to downloading, decrypting and re-encrypting the data with reduced number of calculations under a new access policy. The key is only changed by the data owners, and the key is redirected to the cloud where the data owner can modify the access policy accordingly. In spite of the fact that this paradigm ensures scalability, it is prone to more collusions compared to the previous methods.

To encompass a distributed environment, regular access control system was modified to accommodate a distributed access control system for large data in the cloud that is built on a multi-authority CPABE, with the restriction that no single user can encode/decode the data on their own authority. In this distributed access control model, the server maintains the keys and encrypted data with themselves unless it is requested by the client. The data usually comprises of a private key and certificate. The system also employs a revocation procedure in which the previously revoked clients alone can assist the process. To complete the process of decoding, the client requires a token for the server to decrypt it completely without any interference. But this ultimately compromises the distributed model where the single point of authentication token is used resulting in loss of scalability. This is in contrast with the obfuscation approach [3–6] which also affects the scalability security requirement for IoT.

Pandemic management using latest technologies has been the most needed and many studies have been conducted to review the role of technologies such as AI, 5G, Machine Learning and Blockchain in its aid [7]. There have also been uses of edge and fog computing and deep learning to manage and analyze the large data due to variants of COVID-19 since its inception in 2019 for better healthcare and pandemic management. A novel combination of AI and edge computing for managing the medical information-based contagion of COVID-19 has been discussed by Hussain [8]. Furthermore, the employing Blockchain for supply chain and security requirements of COVID-19 pandemic has also been globally increasing. We will firstly focus on the authentication security requirement for pandemic management in the following.

Tan et al. [9] developed a homomorphic authentication contact tracing system which provided infection monitoring and healthcare management for passengers of their high-mobility transportation infrastructure during these pandemic times. This system was especially developed for "cloud-assisted VANETs" which contains separate modules for data collection and acquisition of healthcare data. The system also employs a decentralized Blockchain vehicle registration between edge and its cloud unit for infection monitoring of a specified vehicle and its passengers. The use of Blockchain in these systems also ensure anonymity and privacy requirement of their passengers. Huang et al. [10] made a contribution in this pandemic scenario by developing a mobile application for tracking movement and finding the persons who came in contact with the COVID-19 infected individual.

A two-way authentication system, entitled HARCI, was proposed by Alladi et al. [11] for use between multiple entities in pandemic scenario applications. The framework was developed in collaboration with the University of California in Berkeley. The HARCI IoT framework is built in three levels containing server node, sink nodes and patient nodes with minimal resources. As a result, the HARCI framework is capable of ensuring authentication between the entities using session keys for every level. "Mao-Boyd logic security proof" has been used to demonstrate that the HARCI framework can assure session key uniqueness while also resisting all attacks.

Later, an ECC cryptographic-based authentication system [12] was developed for wireless sensor networks pertaining to pandemic management. The system is developed to withstand varying attacks namely social engineering attacks, denial of service attacks, brute force attacks, impersonation attacks and many more using "formal security analysis" techniques. In addition, the suggested architecture ensures that users' identities are kept anonymous and that they can communicate with one another. A cloud-based user authentication framework [13] was developed to ensure the authenticity of the healthcare data where the patient node and healthcare sensor node mutually authenticate each other using secret session keys. These session keys are also used for future communication between the two entities. The framework is based on the "ROR model and the AVISPA tool" which had proved itself efficient against brute force and insider attacks.

Wireless sensor networks for these pandemic medical infrastructures often necessitate lightweight device that requires minimal resources and limited power and memory and low computational capability. It is recommended to create a wireless body network by implanting healthcare sensor devices on a patient's body. These healthcare sensors include body temperature sensor, ECG monitoring, SPO2 sensors, blood pressure sensors and many more in requisition with pandemic needs. Later, a smart card-based authentication framework was developed for COVID-19 pandemic management by Das et al. [14]. This framework also made use of the "BAN Logic and the AVISPA tool" which ensures better anonymity than the previous authentication framework against the social engineering attacks, denial of service attacks, brute force attacks, impersonation attacks.

Biometric authentication is an age old technique of using the person's biological traits such as eye, ear and fingerprint to identify and verify that person's identification. These varies from the traditional password-based credential authentication and which involves comparison of biological recorded data with the previously acquired biometrics data to authenticate a person's identification as no two people can be same across the globe.

A guaranteed key agreement approach was developed by Zhang et al. [15] which may meet the security requirements and statistics of healthcare applications. The healthcare system with guaranteed key agreement consists of healthcare aid server is responsible for ensuring that the user's fitness is maintained at all times. A special gold function is used to associate a random character unit with a biometric template to prevent the server from recognizing the template and thus accessing it. Furthermore, more security is ensured by using hash or bio functions and essential chat functions to prevent them from being discovered. When faced with pandemic conditions like COVID-19, this allows the healthcare aid server to acquire and compare the registered biometric features in an aircraft without having to save and obtain accurate numbers. Random numbers are used to protect biometric templates stored on archives such as smart cards and websites, resulting in more security for the users to access the biometric templates actual value without any hindrance.

CDHS, a "cross-domain handshake solution", was proposed in [16] for establishing a safe and secure communication among the participants in their healthcare network, which they applied to the healthcare industry. In order to secure this pandemic scenario, two patients who are enrolled at different healthcare institutions, respectively, can authenticate each other mutually exclusive and employs "elliptic curve cryptography" encryption for establishing sage communication between the two patients suing session key. Furthermore, using the "random oracle model", it was demonstrated that the CDHS solution was safe even though the "Inversion Computational Diffie-Hellman (ICDH)" issue was assumed to be intractably difficult.

Masud et al. [17] developed a secure anonymous key for securing the management of Medical devices in IoT COVID-19 healthcare management. Several cryptographic approaches are included in the proposed strategy which is an integration of hash functions, random nonce and mathematical XOR operations and unclonable functions. According to the results of the experiments, the proposed healthcare approach is capable of protecting communication from adversarial threats by stabling secure sessions with session key. The system also has increased its level of resistance against cloning attacks, manipulation, communication channel threats and assaults, amongst other attacks.

PPDP [18], a "privacy-preserving disease prediction system" was proposed for protecting cloud management in Healthcare systems, which was later adopted to battle COVID-19. This system can be used to give an identity privacy-preserving solution. The proposed framework requires the collection of the historical medical data from COVID-19 patients, which is then encrypted, transferred and stored to a cloud server. From there, the historical data is used to build "prediction patterns" using the "Perceptron Single-Layer learning algorithm" while maintaining patient identity anonymity and confidentiality. From these prediction models, it is possible to determine the illness risk for medical data in the future.

A prediction-based pre-diagnosis system was proposed by Zhu et al. [19] for ensuring safe healthcare systems that were based on the "nonlinear kernel support vector machine". The technique was dubbed eDiag, and it was meant to protect patient privacy. The proposed scheme makes it possible to maintain vital personal health data in a completely anonymous manner in the first stage of pre-diagnosis. Additionally, the system makes use of "polynomial aggregation and lightweight multi-party random masking techniques" to ensure that personal information is protected. Later, a patient-centric privacy-preserving framework was developed [20] to acquire and analyze the risk of occurrence of disease consequently. The proposed framework makes use of a naive Bayesian classifier to analyze the historical healthcare data of the patients without disclosing the current diagnosis healthcare data of the patients. These classifiers enable us to predict the risk of occurrence of disease in genetic and other medical scenario and enable the patients to choose their medical treatment earlier and thereby enhancing the success of medical treatment. As a result, the "additive homomorphic proxy aggregation" approach is used as a way of enhancing the protection of healthcare data anonymity for the patients.

8.3 IoT in Pandemic Management

In order to provide a comprehensive picture of intelligent IoT-enabled structures to combat COVID-19 outbreaks, we study the various conditions that are influencing the use of IoT smart devices and technologies. As shown in the following paragraphs, our study focuses on role of contact tracing application, smart hospitals and other IoT devices in pandemic management as shown in Figure 8.1.

IoMT, which integrates a wide variety of devices (medical or standard) into a variety of health scales and heath care networks, is one of the major resources of Healthcare 4.0 [21]. IoMT can also be defined in different way in literature. For example, "The Internet of Things (IoMT) selects high-level medical devices that are enabled to connect and integrate to high-quality health networks to improve patient health" [22]. These healthcare networks are also involved in healthcare-related processes and medical data management.

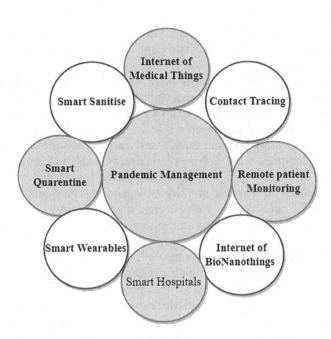

FIGURE 8.1
Pandemic management using IoT and Big Data.

8.3.1 Role of Big Data in Pandemic Management

Big Data can be utilized on the internet to track progress and public pandemic worries, predict pandemic trends and issue early warnings in general public health situations. COVID-19 symptoms vary; however, research into the medical characteristics and signs of COVID-19 positive patients has yet to be completed. One of COVID-190s biggest challenges was ensuring enough safety in the face of a pandemic in the transportation industry. The importance of Big Data technology as a tool to discover virus-prone locations was investigated as one of the five important technological achievements. When COVID-19 is suspected, the diagnosis is made by RT PCR. The test results can take anywhere from 24 hours to several days, depending on the circumstances. As a result, the number of suspected COVID-19 cases has surpassed the current diagnostic capacity in a number of nations. As a result, a number of researchers have developed new methods for detecting COVID-19 infection. In the clinic, a variety of clinical symptoms and signs were obtained and tested, including chest computer tomography, therapeutic measures and medical records. The data was evaluated, and the results were identical to those published by reference for the most common fever and dry cough symptoms.

8.3.2 Role of Contact Tracing Applications in Pandemic Management

Bluetooth has been widely debated as a wireless digital technology for tracking digital contacts, and it has been featured in many legacy healthcare applications for tracking patients. Apple and Google have released iOS and Android developer tools that include links to system application programming interfaces and technologies to make contact tracking much faster and easier. The basic basis for Bluetooth tracking is to find and store Bluetooth signals sent by other nearby devices. With the app on IoMT devices, Bluetooth contact signals are loaded and stored in remote clouds. IoMT device contacts will be detected and notified using Bluetooth signals recorded in COVID-19 certified mode. Although tracking a Bluetooth contact may be visually impaired, the accuracy of detecting IoMT communication devices from a distance between devices is a major concern. Consequently, Bluetooth is also a "peer-to-peer technology", the same cannot be employed for long distance tracking efficiently. Yet, there have been studies that have used AI techniques to employs Bluetooth for contact tracking by increasing the accuracy of distance between peers using Bluetooth Low Energy (BLE) beacons.

Developed countries across the globe have made official COVID-19 contact tracking applications available for various smartphones. "Australia (COVIDSafe), Austria (Stopp Corona), Canada (COVID Alert), Denmark (Smittestop), Finland (Koronavilkku), France (StopCovid), Germany (Corona-Warn-App), China (LeaveHomeSafe), Italy (Immuni), Singapore (TraceTogether), Arogya Setu (India)" and other countries have implemented contact tracking applications for the use of pandemic management and their citizen's knowledge about the affected individuals with anonymity. "COVID Symptom Tracker, HowWeFeel, COVID-19 Symptoms & Social Distancing Web Survey, a global COVID-19 survey, and Beiwe" [23] are some of the tools and methodologies used by these government healthcare workers to curb the spread of COVID-19 pandemic.

For example, the COVID-SAFE [24] tool is designed for healthcare and body distance monitoring using IoT, smartphone application and minimal data analytics intelligence. In addition to Bluetooth, many other NFC-based mobile devices and wireless technologies such as 5G, Wi-Fi, GPS and LoRa have also been used in contact tacking applications. These technologies have a deficit of lower data accuracy and signal qualities which are a great hindrance considering the current pandemic scenario. Consequently, machine learning and AI-based techniques similar to counterproductive production networks can be employed [25] to enhance the quality of signal data and, as a result, improve the accuracy of local performance and position. Fog and Edge cloud computing platform can be employed as a platform, which allows machine learning algorithms and AI techniques to work faster and data privacy issues to be directed at a particular level.

"COVID-19 branding apps, COVID-19 contact tracking apps, COVID-19 health monitoring apps, COVID-19 research apps, telemedicine apps, and social networking apps" are some of the COVID-19 pandemic mobile applications available to the common public and healthcare workers to track the COVID-19 infected individual. These applications serve a variety of purposes. COVID-19 branding apps, for example, can help users track their health, while "COVID-19 tracking apps (e.g., COVIDWISE (Virginia) and GuideSafe (Alabama))" can help healthcare workers to identify the potential victims, who have been in close contact with infected individual. Physicians may use the data acquired by these COVID-19 health monitoring applications to monitor the health of their patients remotely and contactless. To overcome this pandemic hurdle, research organizations have been using these applications to track and investigate the virus and its effects. Telemedicine applications (such as Practo, DoctorLive and Apollo247) can be used by healthcare professionals to facilitate video communication between doctors and physicians and their patients. These applications if integrated with the existing smart city devices and 5G networks with Internet of Things can provide better results.

8.3.3 Role of Other IoT Devices in Pandemic Management

Internet of Things (IoT) is a promising network for vaccination chains of vaccines, health management, drones for delivery, Remote Patient Monitoring (RPM) and identification and fighting infectious diseases such as COVID-19 [26–29]. In healthcare systems, the function of the IoT is to fight pandemics such as the Healthcare Nerve layer, the computer cloud layer and the cloud computing cloud are the three phases that make up COVID-19. Healthcare Nerve layer comprises COVID-19 infected patients, healthcare workers and monitoring wearable devices [30]. These devices monitor the patients progress and communicate the deviations to the healthcare workers periodically and the doctors can uses live video conferencing for less severe patients and employs social distancing to avoid contacting the disease themselves [31, 32].

The concept of internet bio-nanothings (IoBNT) analyzes how the internet and biological cells are connected. As noted by Akyildiz et al. [33], IoBNT also aims to increase the popularity of biomedical research to improve human health and quality of human life. In terms of diagnosis and treatment, IoT technology provides better results and patient monitoring with telemedicine, reduces direct contact with physicians and hospital stays. As a result, medicines and equipment are properly regulated and controlled at low cost due to IoT devices connected to the IoT network.

8.3.4 Role of IoT Empowered Smart Hospitals

Today, smart hospitals face many challenges, including a lack of equipment such as personal protective equipment (PPE) and patient room support, such as ventilators, thermometers, disinfectants, cleaning supplies a shortage of hospital supplies namely physicians, infirmary nurses in response to the COVID-19 pandemic. To address these issues, we are introducing a state-of-the-art hospital utilization that uses IoT technology to expand current infrastructure and enable coronavirus patients to receive integrated care. The literature also looked at a few aspects of the wise hospital vision [34–37]. RFID sensors can be used for supply-chain management of all the entities in a smart hospital. Commonly, RFID sensors are used to track the location of entities and patients inside the hospital. We can also employ additional IoT sensors, and technologies such as Blockchain and AI can help supply-chain management and patient management. Intelligent ventilators, for example, may communicate with patients via smart wrist straps with integrated nerves and respond based on patient physiological features. In addition, Healthcare workers may use IoMT devices with AR technology to monitor patients virtually and make quick decisions. AR technology can also assist the patients with physiotherapeutic and mental health support. Independent robots can also assist in healthcare systems by acting as an automated caregiver taking the roles and responsibilities of monitoring the patients' vitals and food care supply. Surveillance cameras can be modified to use for monitoring the patient vitals in addition to keep track the supply chain management in smart hospitals. The "UV-C (ultraviolet-C) or UVD autonomous cell robot" can disinfect the hard surfaces of rooms and any infected region using "ultraviolet light and Hydrogen Peroxide Vapor (HPV)" and can navigate without any hindrance to the patients by the use of random disinfection algorithm.

RPM is a popular scheme introduced to monitor COVID-19 infected individuals to monitor remotely and provide healthcare solutions remotely with no contact. Companies like Practo and many other telemedicine platforms have been providing access to physicians and telemedicine in these pandemic times and can also adopt RPM for COVID-19 infected individuals. This can enhance the patient–doctor communication and tend to patients with severe need physically and reduce the workload of doctors to a greater extent. Wearable devices for COVID-19 can remotely measure the temperature, blood pressure, blood oxygen level, heart rate, blood glucose level and ECG if necessary.

The collected data is then transferred to a Web Server with the aid of "Message Queuing Telemetry Transport (MQTT) or the Hypertext Transfer Protocol (HTTP) protocol" and stored securely in hybrid cloud using encryption and session keys. The infected individuals may request the physician for COVID-19 test wherein which the physician may employ machine learning and deep learning techniques to predict the occurrence and spread of COVID-19. For example, if a patient's SpO2 level falls below 85%, hospital physicians will receive notifications alerting the critical nature of the patient and thereby ensures faster care for their patients. IoT devices and monitoring application provide patients with built-in repair tips for additional instructions and steps to take at home. The patient may maintain anonymity and also ensures to whom he/she wishes to share their COVID-19 test results ensuring privacy.

Due to security and privacy considerations, he may also invite family members or friends to share his PGHD with various levels of access control. In general, a patient may refuse to disclose their data due to privacy concerns, and in such cases, healthcare providers will send data to the gate device (e.g., patient phone). Apps can also send alerts to a patient depending on the level of limitation, which is especially helpful if the patient's health is at stake. Some warnings, however, may be false ideas, and congested sensory networks, in conjunction with medical equipment, may help to reduce the misconceptions. These types of confusion in patient data can also be avoided by using classification methods such as the "Hidden Markov Model (HMM)" [38]. This model may be used to identify false alarms in the cloud or gate.

To ensure the safety of the healthcare workers who are testing the COVID-19 infected individuals and reduce the expense toward PPE Kit, a BOSTON Hospital designed a testing boot especially for COVID-19 testing with many integrated namely "large infrared body temperature sensor, a non-contact oxygen level sensor with the Red, Green, and Blue (RGB) camera, RFID scanners, and smart AI-assisted cameras". The testing booth allows entry of one individual at a time and the individual is identified by RFID tag or Face recognition using the camera [39]. The testing booth also contains all necessary sensors to mark the infected individual's symptoms. Once the test result is confirmed, it is informed to the infected individual through phone messages or wearable device notifications.

In this section, we will look at other contact tracing and IoT applications with perspective of E-Health. In an emergency, a smart ambulance may provide real assistants to help patients and determine the best routes to the hospital using machine learning services. The smart hospital parking system will automatically scan the driver's license plate and use patient data to identify COVID-19 patients, as well as to choose a dedicated parking space. To prevent inclusion in a single location, patient data is stored in multiple storage cloud solutions. To avoid the single point of failure and to ensure a decentralized management, Blockchain technology can be used to protect patient's data and provide a state-of-the-art threat detection model [39].

Based on the severity of COVID-19, some patients are at risk large enough to be isolated, but not large enough to require continuous care and hospital treatment. They should follow the practice of isolation and stay at home; however, other family members may find it difficult to defend themselves in the same area. In this case, hospital doctors, family members and local officials with varying degrees of need for accreditation should be able to monitor the patient easily. Patients' activities can be monitored at home using both wearable and non-wearable materials [40].

In Smart isolation, a certain level of wearable devices can be attached to the patient's body and some devices and sensors can be strategically placed in the patient's room.

"Visual-based devices (e.g., IR cameras, RGB cameras, and deep cameras), medical devices (e.g., oximeter, scale, and thermometer), and network-based sensor devices" are compressed by three types of wearable devices "(e.g. door sensor, light sensor, and motion sensor)". "Wear items (such as a smart watch, smart mask, and microphone)" can be used by the patient and family members to keep track of their priorities on a regular basis. These data from "Bluetooth Low Energy (BLE) or Wi-Fi-based" devices will be acquired and shared with healthcare workers for further analysis.

Li presented an intelligent cleaning condition [41], which uses IoT technology for intelligent sewage disposal to prevent COVID-19. For example, as the user approaches the house from the driveway, the front door opens using a fine face recognition system and the user enters the house. The device identifies the user's presence and automatically to the home Wi-Fi network for the user. The smart devices of the home will automatically have unlocked and require contactless use with help of speech recognition. The user may place the phone, keys and any other miscellaneous items in a UV light box for cleaning the device and materials. Hand-operated automation can be used at home to prevent transmission of the virus. Voice control gadgets can be used at home. Air-disinfecting systems for clothing and footwear are another future technology that may be beneficial. The closets and baskets they wash can also benefit from this method. Steam may be used to clean smart cars.

During the pandemic, many countries use telephone data to track civilian travel. In addition, some cities and corporations are employing unmanned aerial vehicle (UAV) to monitor the unnecessary moment among its citizens during lockdown period. These IOT-based smart cities are defined by intelligent infrastructure designed to work efficiently to prevent the coronavirus pandemic and save lives. Procedures established by government agencies and provinces should be followed by smart infrastructure. In Nice, Spain, police used drones to force the silence and to disseminate information, which read, "Please check the security measures," including speakers. During the coronavirus pandemic in Handan, Hebei province, China, UAV also sprayed an antibiotic in an infected village. AI-assisted AI aircraft are used in smart cities to gain a distance of up to six meters between people using sensors for measuring mono/stereo distance. There are other types of sensors to measure distance; to measure the social distance between individuals, the mono sensors which have higher accuracy and low cost are fitted in many regions. Certain corporations and government institutions have employed these UAVs for identifying proper face mask of the individual. Many drones monitor residents in smart cities, and algorithms to avoid obstacles and Location Simultaneous Mapping (SLAM) can be used to avoid collisions. As a result, each drone can keep an eye on its location based on its own experience. However, on the computer side of the edge, a total number of parameters from nearby urban areas are collected for group decision-making.

8.4 Security and Privacy Requirements of IoT in Pandemic Management

While employing IOT smart devices for pandemic management, the major concern and hurdle is its security and privacy concerns in its applications and infrastructure management. It is still difficult in the healthcare business to protect connected medical devices and maintain the privacy of users. The COVID-19 infected individuals while testing for

COVID-19 has to give away certain personal and medical information to the testing center which is stored in cloud infrastructure. These data is then in turn shared with the healthcare workers and government to monitor the spread and count disclosure. Separate privacy [42] and data encryption methods [43] (pseudonymize [44], specify name [45]) can be used to protect user identity and privacy. These solutions, however, have their own set of obstacles. For IoT and data-driven applications to be widely used in the prevention, monitoring and reduction of COVID-19, it is important to protect the privacy of users when installing them.

Blockchain with addition to Internet of Things (IoT) can address the security and privacy issues associated with patient anonymity and secure health data. Blockchain is a distributed information platform and immutable by using hash functions, which is flawless and naturally irreplaceable. Data embedded in the Blockchain network is organized in blocks mainly. Each block has a hash value attached to it, which also applies to the previous block, ensuring a recurring connection between them. In an infidelity distribution system, Blockchain provides a consistent, transparent, secure and readable ledger to ensure the integrity and tracking of information/assets throughout their life cycle.

In addition to the chain-link data structure, compact (CM) method is essential for obtaining unique blockchain benefits. CM ensures that transactions are ordered in a clear and consistent manner, as well as blockchain integrity and consistency between geographically dispersed nodes. Performance overruns, delays, node measurements, security level and other aspects of blockchain system performance are largely determined by CM. As a result, many blockchain CMs are proposed based on application conditions and operating requirements. Network performance, delays, storage and durability are all important factors to consider when choosing a CM in an IoT contact tracing application such as BeepTrace [46]. Work Proof (PoW), Stake Proof (PoS) and DAG-based CMs are some of the most commonly used CMs. To overcome the critical difficulty of maintaining confidentiality in digital communication, the COVID-19 pandemic, BeepTrace, is provided with a Blockchain solution. To undermine geodata in user identification, Blockchains are used for patient and physician authentication to access the healthcare data. In terms of security, privacy, battery life and cover, our numerical findings suggest that the proposed BeepTrace is clearly a success. In addition to access control, data flow between the various components of a cloud-enabled IoT platform should be protected from unauthorized access and modification. As a result, Attribute-based Communication Control (ABCC) [47] was born. In order to preserve passwords, a different lightweight authentication method is based on intelligent card and biometric identification [48]. In addition, a distributed computer run by flexible trust systems is needed to engage more cloud and collaboration across a wide range of connected communities. Further research into Blockchain technology [49] is needed to establish a reliable foundation between the various cloud-based systems. Advanced access control systems that use an integrated approach (which includes the benefits of multiple access control models) should be developed to give users access to different levels.

User data privacy issues arise as a result of the recent development of contact tracking systems and will continue to do so in the future. It is recommended to balance privacy and health of users, the need to protect privacy and build contact tracking tools that recognize privacy. Improper machine learning [50], which allows attackers to violate user data and privacy, poses a threat to these and other AI-assisted systems and services. Separate privacy [42, 51] can be used to add audio to data sets for protection. The literature [52–54] discusses in more detail the use of Electronic Health Records in cloud and prior issues.

8.5 Conclusion and Future Focus

The future of healthcare is expected to be digitally focused, which necessitates the recognition of the importance of digital integration in the sector with a pandemic recovery strategy. With the coming pandemic, long-term relationships between professional businesses and hospitals are needed. As a result, digital technology should be used to improve COVID-19 and other infectious diseases, response, resilience, mitigation and future readiness. To combat infectious diseases, digital technology offers great promise. However, privacy concerns prevent the use of such technologies, leading to potentially dangerous situations.

Moreover, relying on a single solution is not enough due to the wide variety of data types and privacy considerations. Our research reflects the future of privacy research, considering the potential benefits of technological advancement. Due to privacy concerns, the limitations shown in Figure 8.1 reduce the value of the digital solution and make it less efficient.

The idea of using cell phones to fight the pandemic is growing worldwide. This has made it possible to track contact, proximity analysis and alerts to avoid high-risk areas. In addition, it has removed most of the privacy issues that may arise when data is stored on a central server. This practice of making multiple calculations on the client side is predicted to continue in the future. As a result, privacy analysis in this context, as well as the proper way to store sensitive information on the user's mobile devices, will be an interesting topic for research.

References

1. S. Fugkeaw, and H. Sato (2015, November). Privacy-preserving access control model for big data cloud. In *2015 International Computer Science and Engineering Conference (ICSEC)* (pp. 1–6). IEEE.
2. K. Yang, X. Jia, K. Ren, R. Xie, and L. Huang (2014, April). Enabling efficient access control with dynamic policy updating for big data in the cloud. In *IEEE INFOCOM 2014-IEEE Conference on Computer Communications* (pp. 2013–2021). IEEE.
3. M. Xiao, M. Wang, X. Liu, and J. Sun (2015, April). Efficient distributed access control for big data in clouds. In *2015 IEEE Conference on Computer Communications Workshops (INFOCOM WKSHPS)* (pp. 202–207). IEEE.
4. M. Elkhodr, S. Shahrestani, and H. Cheung (2013). A contextual-adaptive disclosure agent for general devices in the internet of things. In *Proceedings of the Conference on Local Computer Networks Workshops*, Australia.
5. I. Ullah, and M. A. Shah (2016, September). A novel model for preserving location privacy in Internet of Things. In *2016 22nd International Conference on Automation and Computing (ICAC)* (pp. 542–547). IEEE.
6. J. Sathishkumar, and D. R. Patel (2016, January). Enhanced location privacy algorithm for wireless sensor network in internet of things. In *2016 International Conference on Internet of Things and Applications (IOTA)* (pp. 208–212). IEEE.
7. V. Chamola, V. Hassija, V. Gupta, and M. Guizani, "A comprehensive review of the COVID-19 pandemic and the role of IoT, drones, AI, blockchain, and 5G in managing its impact," *IEEE Access*, vol. 8, pp. 90225–90265, 2020.
8. A. A. Hussain, O. Bouachir, F. Al-Turjman, and M. Aloqaily, "AI techniques for COVID-19," *IEEE Access*, vol. 8, pp. 128776–128795, July 2020.

9. H. W. Tan, P. Kim, and I. Chung, "Practical homomorphic authentication in cloud-assisted VANETs with blockchain-based healthcare monitoring for pandemic control," *Electronics*, vol. 9, no. 10, p. 1683, October 2020.

10. P. Huang, L. K. Guo, M. Li, and Y. G. Fang, "Practical privacy preserving ECG-based authentication for IoT-based healthcare," *IEEE Internet Things Journal*, vol. 6, no. 5, pp. 9200–9210, July 2019.

11. T. Alladi, and V. Chamola, "HARCI: A two-way authentication protocol for three entity healthcare IoT networks," *IEEE Journal on Selected Areas in Communications*, vol. 39, no. 2, pp. 361–369, February 2021.

12. S. Challa, A. K. Das, V. Odelu, N. Kumar, S. Kumari, M. K. Khan, and A. V. Vasilakos, "An efficient ECC-based provably secure threefactor user authentication and key agreement protocol for wireless healthcare sensor networks," *Computers & Electrical Engineering*, vol. 69, pp. 534–554, July 2018.

13. J. Srinivas, A. K. Das, N. Kumar, and J. J. P. C. Rodrigues, "Cloud centric authentication for wearable healthcare monitoring system," *IEEE Transactions on Dependable and Secure Computing*, vol. 17, no. 5, pp. 942–956, September–October 2020.

14. A. K. Das, A. K. Sutrala, V. Odelu, and A. Goswami, "A secure smartcard-based anonymous user authentication scheme for healthcare applications using wireless medical sensor networks," *Wireless Personal Communications*, vol. 94, no. 3, pp. 1899–1933, June 2017.

15 L. P. Zhang, Y. X. Zhang, S. Y. Tang, and H. Luo, "Privacy protection for e-health systems by means of dynamic authentication and threefactor key agreement," *IEEE Transactions on Industrial Electronics*, vol. 65, no. 3, pp. 2795–2805, March 2018.

16. D. B. He, N. Kumar, H. Q. Wang, L. N. Wang, K. K. R. Choo, and A. Vinel, "A provably-secure cross-domain handshake scheme with symptoms-matching for mobile healthcare social network," *IEEE Transactions on Dependable and Secure Computing*, vol. 15, no. 4, pp. 633–645, July 2016.

17. M. Masud, G. S. Gaba, S. Alqahtani, G. Muhammad, B. B. Gupta, P. Kumar, and A. Ghoneim, "A lightweight and robust secure key establishment protocol for internet of medical things in COVID-19 patients care," *IEEE Internet Things Journal*, 2020. DOI: 10.1109/JIOT.2020.3047662.

18. C. Zhang, L. H. Zhu, C. Xu, and R. X. Lu, "PPDP: An efficient and privacy-preserving disease prediction scheme in cloud-based e-healthcare system," *Future Generation Computer Systems*, vol. 79, pp. 16–25, February 2018.

19. H. Zhu, X. X. Liu, R. X. Lu, and H. Li, "Efficient and privacy preserving online medical prediagnosis framework using nonlinear SVM," *IEEE Journal of Biomedical and Health Informatics*, vol. 21, no. 3, pp. 838–850, May 2017.

20. X. M. Liu, R. X. Lu, J. F. Ma, L. Chen, and B. D. Qin, "Privacy preserving patient-centric clinical decision support system on naïve Bayesian classification," *IEEE Journal of Biomedical and Health Informatics*, vol. 20, no. 2, pp. 655–668, March 2016.

21. A. Gatouillat et al., "Internet of medical things: A review of recent contributions dealing with cyber-physical systems in medicine," *IEEE Internet of Things Journal*, vol. 5, no. 5, pp. 3810–3822, 2018.

22. P. Yang et al., "Lifelogging data validation model for Internet of Things enabled personalized healthcare," *IEEE Trans. Systems, Man, and Cybernetics: Systems*, vol. 48, no. 1, pp. 50–64, 2018.

23. Harvard College. Surveys, app. to track COVID-19. [Online]. n.d. Available: https://www.hsph.harvard.edu/coronavirus/covid-19-response-public-health-in-action/surveys-apps-to-track-covid-19/, Accessed on: August. 2, 2021.

24. S. S. Vedaei et al., "COVIDSAFE: An IoT-based system for automated health monitoring and surveillance in post-pandemic life," *IEEE Access*, vol. 8, p. 188538, 2020.

25. V. Chamola et al., "Disaster and pandemic management using machine learning: A survey," *IEEE Internet of Things Journal*, vol. 8, no. 21, pp. 16047–16071, 2020.

26. A. H. M. Aman, W. H. Hassan, S. Sameen, Z. S. Attarbashi, M. Alizadeh, and L. A. Latiff, "IoMT amid COVID-19 pandemic: Application, architecture, technology, and security," *Journal of Network and Computer Applications*, vol. 174, Article No. 102886, January 2021.

27. M. Kolhar, F. Al-Turjman, A. Alameen, and M. M. Abualhaj, "A three layered decentralized IoT biometric architecture for city lockdown during COVID-19 outbreak," *IEEE Access*, vol. 8, pp. 163608–163617, September 2020.

28. I. Ahmed, A. Ahmad, and G. Jeon, "An IoT based deep learning framework for early assessment of COVID-19," *IEEE Internet Things Journal*, 2020. DOI: 10.1109/JIOT.2020.3034074.

29. Z. Fadlullah, M. M. Fouda, A. S. K. Pathan, N. Nasser, A. Benslimane, and Y. D. Lin, "Smart IoT solutions for combating the COVID-19 pandemic," *IEEE Internet of Things Magazine*, vol. 3, no. 3, pp. 10–11, October 2020.

30. S. Misra, P. K. Deb, N. Koppala, A. Mukherjee, and S. W. Mao, "SNAV: Safety-aware IoT navigation tool for avoiding COVID-19 hotspots," *IEEE Internet of Things Magazine*, vol. 8, no. 8, pp. 6975–6982, November 2020.

31. M. Ndiaye, S. S. Oyewobi, A. M. Abu-Mahfouz, G. P. Hancke, A. M. Kurien, and K. Djouani, "IoT in the wake of COVID-19: A survey on contributions, challenges and evolution," *IEEE Access*, vol. 8, pp. 186821–186839, October 2020.

32. A. Roy, F. H. Kumbhar, H. S. Dhillon, N. Saxena, S. Y. Shin, and S. Singh, "Efficient monitoring and contact tracing for COVID-19: A smart IoT-based framework," *IEEE Internet of Things Magazine*, vol. 3, no. 3, pp. 17–23, October 2020.

33. I. F. Akyildiz, M. Ghovanloo, U. Guler, T. Ozkaya-Ahmadov, A. F. Sarioglu, and B. D. Unluturk, "PANACEA: An internet of bionanothings application for early detection and mitigation of infectious diseases," *IEEE Access*, vol. 8, pp. 140512–140523, July 2020.

34. L. Catarinucci, "An IoT-aware architecture for smart healthcare systems," *IEEE Internet of Things Journal*, vol. 2, no. 6, pp. 515–526, 2015.

35. S. Park, and S. Jayaraman, "Enhancing the quality of life through wearable technology," *IEEE Engineering in Medicine and Biology Magazine*, vol. 22, no. 3, pp. 41–48, 2003.

36. L. Catarinucci, D. de Donno, L. Mainetti, L. Palano, L. Patrono, M. L. Stefanizzi, and L. Tarricone, "An IoT-aware architecture for smart healthcare systems," *IEEE Internet of Things Journal*, vol. 2, no. 6, pp. 515–526, 2015.

37. F. Axisa, P. M. Schmitt, C. Gehin, G. Delhomme, E. McAdams, and A. Dittmar, "Flexible technologies and smart clothing for citizen medicine, home healthcare, and disease prevention," *IEEE Transactions on Information Technology in Biomedicine*, vol. 9, no. 3, pp. 325–336, 2005.

38. S. Deep, X. Zheng, C. Karmakar, D. Yu, L. G. Hamey, and J. Jin, "A survey on anomalous behavior detection for elderly care using dense-sensing networks," *IEEE Communications Surveys & Tutorials*, vol. 22, no. 1, pp. 352–370, 2019.

39. S. Li, F. Liu, J. Liang, Z. Cai, and Z. Liang, "Optimization of face recognition system based on azure IoT edge," *CMC-Computers Materials & Continua*, vol. 61, no. 3, pp. 1377–1389, 2019.

40. Y. Li, H. Gao, and Y. Xu, "Special section on big data and service computing," *Intelligent Automation and Soft Computing*, vol. 25, no. 3, pp. 511–512, 2019.

41. S. Rathore, "BlocksecIoTNet: Blockchain-based decentralized security architecture for IoT network," *Journal of Network and Computer Applications*, vol. 143, pp. 167–177, 2019.

42. C. Dwork, and A. Roth, "The algorithmic foundations of differential privacy," *Foundations and Trends® in Theoretical Computer Science*, vol. 9, no. 3–4, pp. 211–407, 2014.

43. E. Vasilomanolakis (2015). On the security and privacy of Internet of Things architectures and systems. *International Workshop on Secure Internet of Things (SIoT)* (pp. 49–57). IEEE.

44. A. Kobsa, and J. Schreck, "Privacy through pseudonymity in user-adaptive systems," *ACM Transactions on Internet Technology (TOIT)*, vol. 3, no. 2, pp. 149–183, 2003.

45. T. M. Truta, and B. Vinay (2006). Privacy protection: p-sensitive k-anonymity property. In *22nd International Conference on Data Engineering Workshops (ICDEW'06)* (pp. 94–94). IEEE.

46. H. Xu, L. Zhang, O. Onireti, Y. Fang, W. J. Buchanan, and M. A. Imran, "BeepTrace: Blockchain-enabled privacy-preserving contact tracing for COVID-19 pandemic and beyond," *IEEE Internet of Things Journal*, vol. 8, no. 5, pp. 3915–3929, 1 March, 2021, DOI: 10.1109/JIOT.2020.3025953.

47. S. Bhatt, and R. Sandhu (2020). ABAC-CC: Attribute-based access control and communication control for Internet of Things. In *Proceedings of the 25th ACM Symposium on Access Control Models and Technologies* (pp. 203–212).

48. L. Kou, Y. Shi, L. Zhang, D. Liu, and Q. Yang, "A lightweight three-factor user authentication protocol for the information perception of iot," *Computers Materials & Continua*, vol. 58, no. 2, pp. 545–565, 2019.
49. B. Tang, H. Kang, J. Fan, Q. Li, and R. Sandhu (2019). IoT passport: A blockchain-based trust framework for collaborative internet-of-things. In *Proceedings of the 24th ACM Symposium on Access Control Models and Technologies* (pp. 83–92).
50. S. G. Finlayson, J. D. Bowers, J. Ito, J. L. Zittrain, A. L. Beam, and I. S. Kohane, "Adversarial attacks on medical machine learning," *Science*, vol. 363, no. 6433, pp. 1287–1289, 2019.
51. A. K. Gautam, V. Sharma, S. Prakash, and M. Gupta, "Improved hybrid intrusion detection system (hids): Mitigating false alarm in cloud computing," *BL Joshi*, vol. 101, 2012.
52. A. F. M. Hani (2014). Development of private cloud storage for medical image research data. In *IEEE International Conference on Computer and Information Sciences* (pp. 1–6).
53. G. Nalinipriya, and R. A. Kumar (2013). Extensive medical data storage with prominent symmetric algorithms on cloud-a protected framework. In *International Conference on Smart Structures and Systems-ICSSS'13* (pp. 171–177). IEEE.
54. E. Zeng, S. Mare, and F. Roesner (2017). End user security and privacy concerns with smart homes. In *Thirteenth Symposium on Usable Privacy and Security ({SOUPS} 2017)* (pp. 65–80). Santa Clara, CA, USA.

9

Sustainable Efficient Solutions for Smart Agriculture: Case study

N. Sudhakar Yadav
VNR Vignana Jyothi Institute of Engineering and Technology, Hyderabad, India

Murali Krishna
PBR Visvodaya Institute of Technology & Science, Nellore, India

I. Sapthami
Visvodaya Engineering College, Kavali, India

Ch. Mallikarjuna Rao
Gokaraju Rangaraju Institute of Engineering and Technology, Hyderabad, India

D. V. Lalita Parameswari
G. Narayanamma Institute of Technology and Science (For Women), Hyderabad, India

CONTENTS

9.1   Introduction 176
     9.1.1   Challenges/Issues in Smart Agriculture 177
9.2   Literature Survey 179
9.3   Proposed Work 182
     9.3.1   Framework 183
          9.3.1.1   IoT -Enabled Smart Farming 183
          9.3.1.2   Smart Soil Selection 184
          9.3.1.3   Smart Irrigation 185
     9.3.2   IoT – BDA Architecture for Smart Farming 185
          9.3.2.1   Processing and Data Architecture.......................186
          9.3.2.2   Random Forest Algorithm 186
     9.3.3   Smart and Sustainable Agriculture 188
9.4   Case Study 188
9.5   Conclusion 197
References..197

DOI: 10.1201/9781003217404-9

9.1 Introduction

The agricultural industry is the most significant sector of the Indian economy. Agriculture in India contributes 18% to the country's gross domestic product (GDP) and offers a means of subsistence for half of the country's population (50%). India is the world's greatest producer of pulses, rice, wheat, spices, and spice goods, and it is also the world's largest exporter of these items. Dairy, meat, poultry, fisheries, and food grains are just a few of the various economic sectors available in India [1]. India has risen to become the world's second-largest producer of fruit and vegetables, and it continues to be one of the world's top three producers in a wide range of agricultural products such as paddy and rice as well as peas and groundnuts, rapeseed as well natural products such as fruits and sugar cane, coffee, and jute as well as cotton and tobacco leaves. On the other hand, on the marketing front, Indian agribusiness continues to face challenges such as low levels of collaboration and incorporation in the business sector, as well as a lack of accurate and helpful information to farmers on a variety of agricultural concerns [2]. To provide substantial results, BDA is coupled with business processes and traditional analytics. BDA may be divided into four categories: descriptive analytics, inquisitive analytics, predictive analytics, and prescriptive analytics, among others. Considering the significance of agriculture as a major source of food, the sector's long-term development is crucial. As a nutshell, a system that preserves and enriches the natural resource base while improving productivity is required for sustainable agriculture. The importance of technologies in smart and sustainable agriculture is illustrated in Figure 9.1.

"Big data analytics" is the technique for examining large volumes of data to uncover hidden patterns and relationships, as well as market trends, customer needs, and other useful information for commercial purposes. In the real world, theoretical outcomes might lead to more successful marketing strategies, additional revenue opportunities, improved agricultural planning, greater performance of producers, competitive advantages over rival societies, and other economic benefits. This generation of agricultural professionals must enhance decision-making processes that can take use of considerable gains in data and information from a variety of various sources, including land, crops, weather, and farm management systems [3] throughout the 21st century. Paper [4] sought to obtain an understanding of agricultural output forecasts via the use of big data analysis while also recognising the socio-economic problems involved. The analysis of this massive quantity

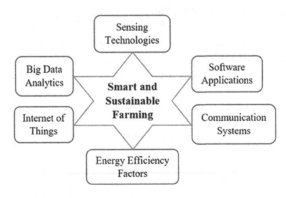

FIGURE 9.1
Role of technologies in smart and sustainable agriculture.

of data is based on K-means clustering techniques, which are used to determine which farming methods are best suited for a certain region's farming practices and to anticipate yields using Apriori algorithms. This useful knowledge was once again made available to farmers in order to improve crop yields and promote organic farming. According to Victoria et al. [5], an analytical framework for a parallel and decentralised feature discovery Hadoop-based neural network in large data sets has been developed. [6] They developed five attribute selection methods by using an artificial neural network system that was built up in the Hadoop YARN cluster. They discover the best attribute selector for the simple identification of acceptable characteristics from wide and high dimensional data sets by integrating the Hadoop binary relational-memory neural network with dependability and mobility to get the best attribute 28 selector. Hadoop's MapReduce representation has been enhanced by Lin [7], who has integrated the Apriori data mining technique into the model. When dealing with many characteristics, the rule scarcity is calculated using algorithms that are often associated rules. The MR-Apriori method that has been presented solves this problem. Paper [8] developed a technique for making effective crop recommendations. The model, which is primarily concerned with agriculture in the Telangana area, makes crop recommendations to farmers via the use of a Nave Bayes MapReduce classifier. Since it may be used to test on other crops, the system is scalable. The optimum time to seed, the best time to grow the plant, and the best time to harvest the plant may all be determined from the yield graphs. A precision agriculture model is offered to help farmers choose which crop to produce based on the circumstances of their fields. Paper [9] have developed simple exponential smoothing, holt's linear smoothing, and holt-exponential winter's smoothing techniques for predicting region-wise rainfall using the MapReduce computing model, which they have implemented using the MapReduce computing model. It is decided to undertake the experimental investigation on two separate datasets. The first is the Indian Rainfall dataset, which contains information on the year, state, and monthly rainfall in millimetres (mm). In the second place, there is the Tamil Nadu state rainfall dataset, which contains information on the year, districts, and monthly rainfall in millimetres. The MSE accuracy metric is computed to verify these approaches. In the study's findings, Holt-Exponential Winter's Smoothing is shown to be the most accurate method for rainfall prediction. Paper [10, 11] built a system that uses historical weather data from an area to analyse the data. The MapReduce and Hadoop methodologies were used to accomplish so. Many key industries that are impacted by the climate, including as agriculture, aviation traffic, water resources, and tourism, may benefit from weather forecasting services. Weather forecasting is a branch of meteorology that is carried out by gathering data from numerous sources that are connected to the current weather conditions such as rainfall, temperature, wind, and fog, among other things. On the Internet of Things (IoT) age, the meteorological department employs a variety of sensors to measure humidity, temperature, and other parameters. The MapReduce technology is employed for the distributed algorithms that are used to properly analyse the meteorological data. The advantage of Map Reduce with Hadoop is that it will speed up the processing of data in an environment where the amount of data is growing by the day [12].

9.1.1 Challenges/Issues in Smart Agriculture

Despite a number of warnings, natural farming has only just begun to take off, impeded by the anti-counterculture stigma attached to it. The studies from the 1990s demonstrate the presence of insect killers and chemical compounds in food intended for people, mostly youngsters. Due to the increasing need for naturally created food, the development of natural farmhouses is being encouraged [13]. The Internet Archive has grown enormously during the decade of 2010, and it continues to develop daily. The internet is now the primary source of information for most people, and it is the most effective platform for selling crops on the market and promoting agricultural operations, as well as for delivering information to farmers directly. Farmers are increasingly using online sites and social media to gather knowledge on growth approaches, agri-market information, and to exchange novel intercropping ideas with other farmers. Many agricultural information websites, such as iKisan.com [14], are hosted and maintained by the Indian government. The Agriculture Ministry announced the commencement of the new "Kisan Call Centers (KCCs)" programme on January 21, 2004. The primary goal of this contact centre is to provide answers to farmers' queries in their own language. Several call centres are in each state, and they are capable of handling calls from any area of the nation. These contact centres are responsible for answering questions pertaining to the four agricultural fields as well as those pertaining to specific industries. Farmers from any region of the state may contact the call centres by dialling toll-free lines, and they can get explanations on any farming-related questions they may have. Kisan call centre operators respond to farmers' concerns or questions as soon as they can by answering their phone calls on the spot. In the case that the operator at the contact centre is unable to respond to the farmer's question promptly, the call will be routed to the agricultural experts [15] for further assistance. By the end of July 2013, the honourable President of India has officially launched the SMS Portal for Farmers. In May 2014, the mKisan SMS Portal was launched in rural regions for the benefit of farmers, who may now get agricultural information, services, and advisories by sending SMS to their registered mobile phone in their local language, as well as by sending SMS to their registered mobile phone. The farmer may register his mobile number by visiting the internet URL http://mkisan.gov.in/wbreg.aspx and following the instructions. Alternatives include visiting a local Common Service Center (CSC) and registering his cell phone number with the aid of a Village Level Entrepreneur (VLE) (VLE). The VLE [16] charges a one-time cost of Rs. 3 for each registration that is completed. Farmers have been able to utilise the Internet to find solutions to agricultural difficulties as the number of people using the Internet and the number of people using smartphones in rural regions has increased. The government, in collaboration with a variety of businesses and non-governmental organisations, has developed mobile app apps for farmers that provide real-time information on the weather, local markets, seeds, and fertilisers, among other things [17]. Additionally, farmers may communicate with and get help from agricultural specialists from all around the nation using the applications. Mobile phone apps have been developed by the Indian government to assist farmers in obtaining crop-related information such as insurance and current agricultural commodity prices throughout the country via their mobile phones. The following issues were detected because of the above. As a result of the continued employment of traditional agricultural practices by many farmers, crop yields are very poor. Agriculture scientists and researchers are investigating novel strategies for obtaining a high yield from farmland, which has attracted the attention of farmers, government officials, and agricultural scientists and researchers Growing high-quality crops is difficult since crop output is mostly dependent on the amount of soil fertility. In addition, it is necessary to identify crops with low nutrient content and to optimise such crops to meet the aim. Crops are often damaged by a shortage of basic nutrients. The application of fertilisers of the appropriate grade is thus essential. It is difficult for farmers to acquire information on their farmland's soil nutrient content, water nutrient content, ground water level, weather condition, and seasonal crop data. Furthermore, people are having difficulty making better decisions based on the information that is accessible to them.

India is an agriculturally oriented nation, with agriculture providing employment for more than half of the population. According to the latest available data, agriculture employed 50% of the Indian labour force and contributed 17–18% to the country's GDP in 2018. It is all the more important that agribusiness be committed to increasing India's national GDP. Agriculture serves as the foundation of the Indian economy [18]. Crop cultivation is now at a low level due to the fact that many farmers are still using traditional agricultural practices. In order to maximise the production from farms, agricultural experts, government officials, and researchers are investigating novel techniques of increasing yields. In order to enhance agricultural yield, the Indian government is focusing its efforts on agriculture via a variety of initiatives. Using the Green Revolution as a foundation, new concepts for individual crops may be offered 2. They are reliant on natural variables such as water, soil, climate change, and so on. The fertility of the soil is the most important factor in crop yield. As a result, it is essential to identify crops with low nutrient content and to increase the amount of nutrients available to such crops. In India, soils have been shown to be deficient in the main nutrients (nitrogen, phosphorus, and potassium). To overcome the deficit, it is essential that the appropriate grade of fertilisers be used in conjunction with a crop's growing conditions. "Soil health" may be defined as a mix of soil's physical, chemical, and biological activity. With relation to agriculture, the primary purpose of any technology is to increase crop output while providing several immediate and long-term advantages to the farmer. The Indian government has taken the required efforts to address this issue in order to ensure that farmers get a fair income via the use of information and communication technology (ICT) [19].

9.2 Literature Survey

Each farmer aspires to make an educated guess about the result of his or her produce based on their previous succeeding history. Although this projection is based on historical data, there may be differences depending on the meteorological environment, insect infestation, and harvest operation schedule. Exact information on the previous performance of crop yield value is required before any inferences can be drawn in the context of agricultural risk management planning. Improvements in agro-ecological and socio-economic conditions are achievable when agricultural decision systems are used in conjunction [20]. Many applications in the agricultural area employ data mining techniques and BDA methodologies, which are used to make predictions based on the data collected. Current methods of predicting agricultural yield rely on statistical analysis, which, however, does not reflect all of the available knowledge representations in full. It has been discovered and tested by scientists all around the globe that there are multiple different statistical approaches applicable to the field of agriculture. Some of them are shown in the following list: Gholap et al. [21] developed a decision tree system to estimate soil fertility in a field setting. It was discovered that the performance of the J48 decision tree method might be improved by collecting data from a private soil testing facility in Pune and applying attribute selection and boosting techniques to it. J48 is a Java 33 implementation of the C4.5 algorithm that is available as free source software. It is a statistical classifier based on the Id3 method, which is often used in machine learning applications. It operates based on the idea of information entropy. Using C4.5, a decision tree is generated, with each node dividing the classes according to the amount of information gained. Using normalised information gain as a splitting criterion, the attribute with the greatest normalised information gain is selected. They projected the amount of soil fertility and classed it as very low, very high, low, high, moderate, and moderately high based on their predictions. Additionally, they increased the accuracy level to 96.73% after using the selection and boosting method, according to the researchers. In the paper [22], the authors developed a novel strategy for soil categorisation and forecasted different characteristics with the aid of a decision support system. With the use of this new decision support system, they were able to keep track of agricultural operations. A quick decision support procedure was obtained via the use of the Coimbatore District dataset and the upgraded C4.5 algorithm, which was developed by the researchers for this work. For the soil data set, paper [23] concentrated on classification and clustering data mining methods, which they found to be effective. The soil information used in this study was obtained from a soil testing facility in Bahadurgarh, Jhajjar, India. In all, there were 49 soil sample instances with a total of ten characteristics. Therefore, soil classifications are concerned with the categorisation of soil into distinct soil classes based on the nutrient percentage present in the soil. Fertiliser recommendations are made in accordance with these classifications (for example, very low, low, medium, and very high). Based on correlation research, Paper [24] has reported on the adaption of Indian farmers to mobile-based agricultural advising systems without considering their social and economic aspects. Furthermore, they have had the convenience of learning crucial information on pest and disease management via the use of a cell phone. For rural residents, a call centre based on a mobile multimedia agricultural advisory system (MMAAS) offers the information they need. Using supervised learning and back propagation neural networks, Ghosh et al. [25] developed a unique technique for the study of soil attributes that was previously unexplored. The investigation of how soil qualities, such as organic matter, vital plant nutrients, and micronutrients, impact crop development, as well as the determination of the optimal degree of connection between these variables, were the primary motivations for this study. Costly and time-consuming direct measurement of these parameters is required. The results demonstrated that the new technique was more accurate in predicting the soil qualities than the previous approach. Data mining classification and clustering methods were used in the investigation of soil resources and soil types carried out by Hemageetha et al. [26], which was conducted in the Salem District. It assists farmers in making informed decisions about which crops are best suited to their land. The soil of the Salem District is a mix of red and black. Using the classification method, the most appropriate crop was selected depending on the soil type and fertility level of the field. Using classification algorithms, Paper [27] has proposed a technique for predicting the soil type that may be used in agriculture. It was discovered that soil data included hidden information via the use of JRip, J48, and the Nave Bayes method. They acquired data from 110 soil test reports from the agricultural soil testing facility in Sattur Block, Virudhunagar District, which they then analysed. They compared the findings to kappa statistics to determine the most appropriate method. The JRip model produced correct findings for the data from the soil test report. According to Paper [28], a new strategy to estimate agricultural production in Tamilnadu, India, that is based on the data mining association rules system has been proposed. It has been decided to use a variety of methodologies for the evaluation of farming and the calculation of crop yield. The authors developed and deployed a model to estimate crop yields based on existing historical data. Paper [29] has investigated the relationships between weather patterns on a wide scale and crop yielding conditions. A total of 35 artificial neural networks (ANNs) have been established as dominant tools for modelling and prediction in order to improve their performance. The Artificial Neural Network (ANN) technique for crop forecasting is used to forecast the correct harvest by sensing the different parameters of the soil as well as environmental parameters such as type of soil, PH, nitrogen, phosphate, potassium, organic carbon, calcium, magnesium, sulphur, manganese, copper, iron, depth, temperature, rainfall, and humidity using the Artificial Neural Network (ANN) technique for crop forecasting (ANN). A method developed by Armstrong et al. [30] aimed to assist farmers in Western Australia by enhancing decision-making in the selection of plant types for different climatic and agricultural settings. Six major components made up the AgMine DSS: modules for data input and mining, statistical analysis, database management, prediction, visualisation, and display of prediction results. The ArcGIS approach was used to visualise weather and other information for two districts (Milling and Wongan Hills), and the results were rather impressive. In this study, we used the Arc Map and the conventional kriging procedure to interpolate annual precipitation and plant yield using spherical semi-variogram models for the years 2008 to 2012. Another problem, according to Patodkar et al. [31], is providing logistics in rural regions without boosting farmer's prices or cash on delivery payments. Each crop type has its own fertiliser policy, which has been recorded. If the temperature has reached the desired temperature range for the crop's sowing date and the temperature has reached the desired temperature range for the crop's sowing date, the farmer will send out notifications about the application of fertiliser, herbicide as scheduled, disease pesticide, and weather alerts. The crop sowing concepts that are required are dependent on the kind of soil and geographic area. The farmer has been provided with the most recent national crop pricing in order to add value to his operation. Using GPS, this system incorporates modern Internet and mobile communications networks to provide safe and efficient farm operations. Paper [32] concentrated on the digital analysis and subsequent decision-making of agribusiness tech-savvy farmers. Farmers in the state who are younger and more educated prefer to utilise the internet to make crop-related choices rather than talking to one another. Agriculture is a specialised industry, and as a result, the m-commerce-related capability for Agri input must be tailored to the individual needs of each site. The most difficult issue to overcome is the inability to comprehend how these m-commerce networks may provide value to the Agri-input distribution routes. Another issue is how to supply logistics in rural regions without boosting the farmer's price or increasing the amount of money spent on payment distributions. Nguyen et al. [33] established an intelligent framework for diverse items that makes use of time series models to estimate short-term market conditions. The system necessitates a number of activities, including the collecting of online selling information, the pre-processing of raw materials, and the usage of an ARIMA design. The efficiency measurements Root Mean Square Error (RMSE) and Mean Absolute Percentage Error (MAPE) have been employed in the technical prediction accuracy study to determine the accuracy of technical predictions. PriceMe's website provided the sales data that was used in this report. They also compared and contrasted the ARIMA system with the Moving Average model, as well (MA). In the instance of the MA system, the anticipated patterns described by a flat line are easily discernible from the actual patterns. The ARIMA system did not do well when it came to projecting long-term trends. McNally and colleagues [34] have presented a method for determining the predictability of the direction of the Bitcoin price in US dollars. The pricing information for Bitcoin was obtained from the Bitcoin price index. It was possible to complete a job with varying degrees of success thanks to the implementation of a Bayesian optimised Recurrent Neural Network (RNN) as well as a Long Short-Term Memory (LSTM) network. The LSTM obtained maximum classification accuracy of 52% and a root mean square error of 8%. In order to compare the deep learning models with the common ARIMA model for time series prediction, a common ARIMA model was introduced. Kaur and colleagues [35] have written on data mining tools and approaches for agricultural applications. When it comes to more contemporary data mining technologies, methodologies such as K-Means clustering, K-Nearest neighbour clustering, Artificial Neural Networks, and Support Vector Machines are used. They are looking at the subject of agricultural price predictions. This has grown into a very serious agricultural problem in recent years, and it can only be resolved with the help of the knowledge currently accessible. The researchers developed relevant information models that aided in the accomplishment of high accuracy and generality in the prediction of price movements. Using different techniques of prediction, Ticlavilca et al. [36] conducted an investigation into agricultural commodity prices. In order to get multiple-time-ahead estimates, we applied a multivariate relevance vector machine that was based on a Bayesian learning machine technique to regression in order to obtain multiple-time-ahead estimates. On the basis of its efficiency, the MVRVM model was compared to the quality of another system that produces many outputs, such as the Artificial Neural Network (ANN) (ANN). The method of bootstrapping is used to test the robustness of the MVRVM and ANN models. Paper [37] has proposed a novel vegetable price prediction model for the Back-propagation Neural Network (BPNN) and Radial Base Neural Function Network (RBNFN) that incorporates the Back-propagation Neural Network (BPNN) and Radial Base Neural Function Network (RBF). The anticipated results of the neural network were compared to the actual outcomes. The RBF neural network is proven to be more effective and dependable than the BPNN in terms of performance and reliability. The tomato value in Coimbatore was modelled, and MATLAB was used to anticipate the models' predictions. Luo et al. [38] investigated of four models to predict the wholesale price of Lentinus edodes in Beijing Xinfadi, and their findings were published in the journal Agricultural Economics. To put it another way, the neural network system BP, the neural network model based on genetic simulations, the neural network model RBF, and an integrated prediction model based on the three systems are all discussed here in detail. They utilised a maximum of 84 papers gathered between 2003 and 2009 to feed into the four prototypes of training and assessment that they developed. 38 The neural network model developed by BP produces the lowest results. The neural network model based on the evolutionary algorithm proved to be more trustworthy than the neural network RBF technique. Shim [39] has advised that the BP neural network be used to estimate the price of vegetables on the market. For example, the tomato information market price in Coimbatore was tracked for three years and the results were calculated using MATLAB. To anticipate the monthly and weekly vegetable market values, they used a neural network as a way of prediction approach in conjunction with the nonlinear time series. A comprehensive error ratio of monthly and weekly vegetable prices forecast was obtained, and the accuracy ratio for the price prediction was analysed as well. As a nutshell, it is clear that the development of smart agriculture is critical, and that sustainability must be prioritised; thus, the demand for a cost-effective approach for smart and sustainable agriculture must be considered.

9.3 Proposed Work

Data mining is a multidisciplinary subject that involves many different disciplines. It is used to extract information from data that has been concealed from view. The enormous volume of data collected across a wide range of areas presents data mining algorithms with remarkable obstacles. When it comes to the efficient use of data mining tools and methods, researchers and practitioners agree that attribute selection is a critical component of the process. A technological concept in feature selection allows for the automated selection of features from a data collection using a mathematical formula. That is the most relevant information for the issue. The characteristic of data is referred to as a feature. Using Attribute Selection (AS), you may choose a small subset from the original attributes by following a criterion, for example, it is a method of picking M attributes from the original set of N attributes, where M is the number of attributes in the original set of N. It is one of the most important and crucial data preparation methods in a variety of fields, including artificial intelligence, data mining, and machine learning, among others. The relevance of an attribute is divided into three categories: high significance, weak significance, and irrelevant. A significant significance in the data set refers to the optimal subset of qualities that are regarded vital for the prediction and cannot be deleted. In order to determine the ideal subsets of qualities for which selection is required, certain requirements known as weak significance must be met. The unnecessary characteristics must be removed since they provide no information that is relevant to the target.

9.3.1 Framework

The proposed framework starts with collecting data with the support of IoT enabled sensors. All the collected values shall be processed and analysed using BDA algorithm. Moreover, it results in a smart framework for smart agriculture especially in India. Big Data and IoT jointly improves the smart agricultural workings in a better way and thereby it maintains a sustained environment green and better for future.

9.3.1.1 *IoT -Enabled Smart Farming*

Smart farming is already there in the market benefiting farmers and vendors in improving the productivity and healthy grains. Smart farming can also be made very effective and energy efficient with the integration of Data analytics algorithms to find the insights and pattern of soils and also to make a farmer to much familiar with the fertilisers to be used at right time and giving required water supply for smart cultivation. In smart farming, IoT/sensor nodes are essential for obtaining real-time data [40, 41]. These components have the ability to make the process more realistic by gathering real-time data from farm fields in order to improve the precision of the agriculture system. The agriculture system becomes more functional by incorporating data analytics and machine learning with IoT. All of these technologies have a wide range of uses in different sectors. Various approaches for farmers are being created in precision farming to keep them up to date on the status of their crops.

The Big Data-IoT-enabled precision model comprises of four phases as shown in the Figure 9.2. The very first element comprises of a large number of sensors/IoT nodes that monitor physical or environmental factors, ground conditions, and plant conditions. For example, a moisture sensor captures soil wetness readings, while a soil nutrient sensor checks the soil's fertility. Sensors those used most widely in smart agriculture are shown in the below Figure 9.3.

There in second stage, we should collect this precise information. Depending on the necessity, we could either store the data locally at the local fog node or send it to the cloud for higher processing and remote monitoring. The analytics techniques are being used in the third stage of the architecture to determine the accuracy of the crop fields. The final phase employs algorithms for prediction, visualisation, control, and alerting, which must be communicated to the farmer for improved cultivation. This data is subsequently passed

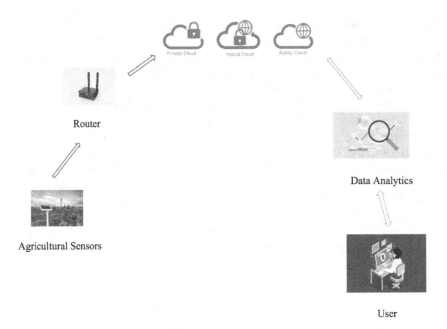

FIGURE 9.2
Big data-IoT-enabled precision agricultural model for smart farming.

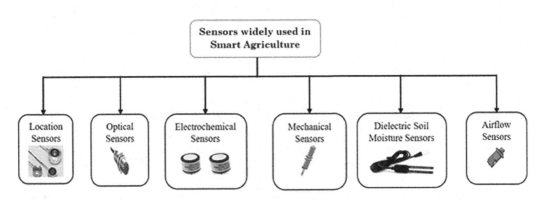

FIGURE 9.3
Sensors most widely used in smart agriculture.

on to the end farmers, who may then decide whether the measurement is below or above the threshold. As a result, they receive a notification to the actuator, which turns on (or off) the watering system, allowing water to flow into the soil. Alternatively, the farmer may need to apply fertilisers such as potassium, nitrogen, and phosphorus to balance the soil's fertility. Any crucial scenario is recognised sensing/predicting and a reaction mechanism is activated using analytics and actuators.

9.3.1.2 Smart Soil Selection

Every agronomic activity is built on the foundation of soil. There is no such thing as a crop without soil. The tummy of flora is thought to be the soil [42]. So assessing the state of the soil is the first and most important step toward the finest agricultural practice. We can examine the physical, biologic, and chemical characteristics of the soil by doing soil testing. Farmers can make informed judgments about their fields based on this information. Precision agriculture is primarily motivated by the desire to produce more from the limited amount of cultivated area farmland. As a result, introducing innovative era technologies is crucial all over the world, as we must satisfy an ever-increasing population with fewer resources.

Soil tests predicated on weather and soil conditions are usually performed in the fall [43]. We may examine the soil nutrients by performing soil tests, which includes information such as fertiliser demands, copping records, soil composition, irrigation rate, and so on. We also have a range of sensor-based techniques to help with soil choices and management. These tools assist us in determining the greatest crop for the best soil. Agrocares Lab-in-a-Box soil testing toolbox is considered to be the ultimate laboratory test toolset [43]. Any farmer can use this to test the soil in their field without having to visit an agriculture facility. Any farmer, according to this toolbox, can analyse roughly 100 samples per day without requiring any laboratory training. It means that roughly 36000 samples may be analysed each year without having to go to a lab. Paper [44] enhanced seed spraying using GPS (global positioning system), sensors, and an autonomous robot based on perception named Agribot. In summary, we can state that current technology is extremely beneficial to every farmer in assisting them in identifying the finest land for the best yield.

9.3.1.3 Smart Irrigation

Irrigating crop fields on a need-to-know basis is one of the most effective ways to increase crop yields while also managing available fresh water resources. Only 0.5% of the available fresh water is used by humanity. Salt water makes up roughly 97% of the accessible water on the Earth, with fresh water accounting for the remaining 3%. Approximately 67% of fresh water is frozen in polar ice caps or glaciers. It means that just a small fraction of fresh water remains unfrozen beneath, leaving only 0.5% for vegetation and fauna to survive. As a result, it is humanity's obligation to preserve water resources. Many researchers are working towards this goal. As a result, traditional irrigation techniques can be managed through the use of emerging technologies such as IoT and Big Data.

We could perhaps set up a Big Data-IoT-based system that monitors soil moisture and alerts us when it's time to irrigate the process. Crop Water Stress Index (CWSI) has been created based on IoT in paper [45], which may be utilised to improve crop efficiency. We deploy the requisite sensors in the required field and gather data, which is then sent to the central processor in the CWSI system. We also collect the data from weather data, using satellite photos, at the central node, and decide whether or not to irrigate the field based on that data. In an essence, we can state that by implementing emerging technology, we will be able to retain the essential moisture for the vegetation while also conserving fresh water supplies.

9.3.2 IoT – BDA Architecture for Smart Farming

The IoT system encompasses the pervasive presence in the ecosystem of a variety of things and devices capable of establishing a dynamic ecosystem via wired and wireless connections as well as preferred communication standards. It reduces human resource costs and provides improved coverage, connectivity, authentic, and monitoring capabilities across multiple sectors. It also opens up an abundance of new prospects in a different domain, as well as a huge possibility for agriculture.

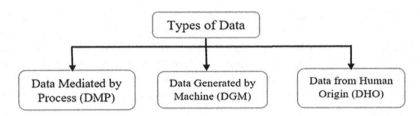

FIGURE 9.4
Types of data produced in the agriculture sector.

Big data has the potential to intervene in SF and overcome the constraints of conventional decision-making and data storage systems, which are incapable of handling large volumes of data. Sensor deployments and analytics, data modelling, and predictive are used in agriculture to mitigate the damage to crops, boost weight gain and feed inbreeding, and maintain the efficacy of agricultural methods in various fields.

The IoT is providing farms with vast amounts of data that can be accessed in real time or in batches. As illustrated in the diagram Figure 9.4, there are mainly three types of data produced in the field of agriculture.

Big Data solutions for Smart Agriculture can only be effective if they are accompanied by a data migration strategy that includes data modelling, computation, storage, and performance, as well as systems management and setup, and an architecture that satisfies all of these necessities as well as the technical constraints of data files and data processing performance for the Smart Agricultural process. While a task is being executed on machine x, at the very least, resource, job, data block migration, and network limitations should be considered in order to ensure that here should be no more RAM used on the machine than what is required for all jobs and slots.

$$\Sigma_-(s=1)^\wedge(m_-d^\wedge s)\Sigma_-(d=1)^\wedge N v_-d^\wedge r \leq m_-x^\wedge r \qquad (9.1)$$

9.3.2.1 Processing and Data Architecture

Developing a Data Model that fits all data processing needs for Smart Farming Analytics is the biggest concern. The architecture of data processing model should ensure the points given in the Figure 9.5.

9.3.2.2 Random Forest Algorithm

Random forest is a supervised learning model that may be applied to both regression and classification problems. The Random Forest algorithm builds decision trees on distinct data samples, predicts data from each subset, and then decides on which option is optimal for the system. This is illustrated in Figure 9.6. The random forest is a technique of ensemble learning that uses several decision trees. Random forests tend to enhance prediction by averaging the impact of numerous decision trees. There are a variety of models for making predictions based on classification data. For binomial variables, one of the most used methods is logistic regression. Support vector machines, naïve Bayes, and k-nearest neighbours are some of the other methods. Random forests thrive in cases where a model contains a high number of features with modest predictive ability individually but significantly stronger predictive ability cumulatively. The data was trained using the bagging approach

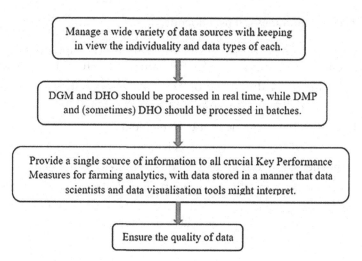

FIGURE 9.5
Architecture of data processing.

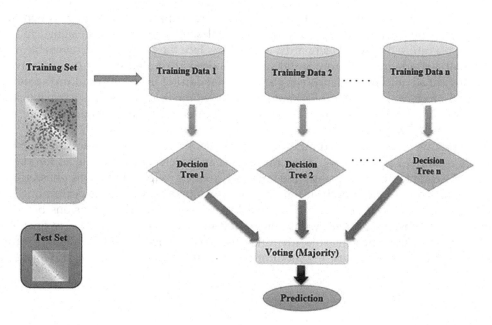

FIGURE 9.6
Random forest technique.

using Random Forest. Essentially, the bagging approach is a combination of researching many models in order to improve the system's final outcome. We employed the Random Forest technique to achieve high accuracy, which provides accuracy that predicts by model and the eventual outcome of prediction in the dataset. In the random forest, a decision tree is created from a data sample, and the trees provide predictions for each family. The best solution is chosen by voting, which improves the model's accuracy. It produces the best outcome for such system.

Random forest classification method pseudocode:

Step 1: Input the pre-processed values of soil set attributes (Phosphorous (P), Nitrogen (N), Temperature (T), Potassium (K), PH value (PH), Humidity (H), Rainfall)

Step 2: Select the feature (i) randomly out of the total feature in the model (m).

Step 3: Choose the feature (i) and compute the node.

Step 4: Compute predictable frequency for each element.

Step 5: Compute observed frequency for each element in a sequential manner.

Step 6: Combine the measured and predicted square values by the predicted frequency.

Step 7: To be the next node, must choose component with the highest weight.

Step 8: Remove the node element and repeat the process until all the relevant elements are found and all the attribute considered are applied.

Step 9: End.

9.3.3 Smart and Sustainable Agriculture

The demand for sustainable agriculture is rapidly rising as a result of environmental factors such as water scarcity, unsustainable agricultural expansion, soil degradation, and carbon emissions. The paradigm of environmental sustainability was designed to help reorganise a future that was in jeopardy. Many areas face the prospect of irreversible environmental degradation. Environmental strain has been attributed to population increase, technological advancements, and rising affluent standard of living. Agriculture's role as a vital food producer necessitates its long-term development. As a result, sustainable agriculture needs a novel system that both protects and enriches the natural resource base while also enhancing productivity.

The objective of the proposed model is to measure the sustainable performance of sustainable farming. There are five categories in this approach. The lowest level for metrics that can be used to calculate environmental practices are group metrics. This phase gives a performance indication for each sub-dimension to evaluate how well it is performing. The third category is sub-dimension phase, which contains eight sub-dimensions based on three main dimensions. Monetary, environmental, and social criteria are used to categorise sustainability. The upper quartile identifies the goal, which is long-term agricultural sustainability. The framework for monitoring sustainability performance is depicted in Figure 9.7.

9.4 Case Study

The concept of machine learning systems to learn and create models for future predictions seems to be well, and for legitimate reason. Agriculture is vital to the world economy. Understanding global agricultural output is critical to tackling food security concerns and mitigating the consequences of climate change as the population continues to expand. This chapter proposes an efficient technique to assist in smart agriculture and helps the farmers to get a better yield and good profit. The proposed technique has been shown with a

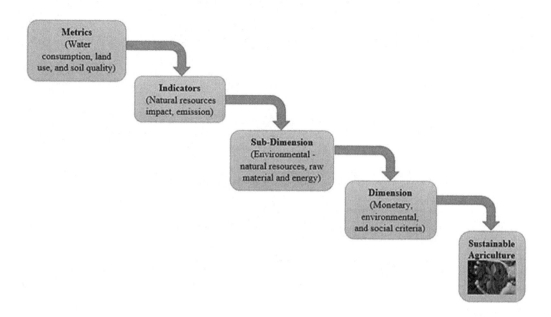

FIGURE 9.7
Framework for monitoring sustainability performance.

case study in this section. The framework has been deployed in one of Andhra Pradesh's farmlands. The efficiency of the proposed method is discussed along with its limitations.

The system gets input data from multiple sensors and pre-processes it, after which training data is fed into the system using machine learning algorithms. The random forest algorithm is being used to obtain at a better choice based on the quality of the data. The farmer would be informed about the best plantation for that area, as well as which crop may deliver the best production with such a soil type. Smart irrigation is also an element of this scheme, in which water will be delivered when needed and water waste will be prevented. The proposed technique has the ability to process large amounts of data. The framework is more effective for sustainable farming since this is based on the convergence of BDA and the IoT.

The numerous crops that provide higher yields in the considered state are depicted in the diagram Figure 9.8.

The graph Figure 9.9 illustrates the crop production by district. East Godavari and West Godavari provide higher yields throughout the year, and some crops are observed to achieve higher yields in certain areas.

It is possible to construct a mathematical model using regression analysis in which the predicted value of a dependent variable Q (represented in matrix form as q_i) is calculated in terms of the value of an independent variable (or vector of independent variables) P, as shown in the equation below:

$$q_i = \beta_0 + \beta_1 p_i + \beta_2 p_i^2 + \ldots + \beta_n p_i^n + \mu_i, \text{where i} = 1, 2, .., n \tag{9.2}$$

where q_ii denotes the *i*th dependent variable value, β_0 denotes the intercept, β_i denotes the *i*th angular coefficient, and p_i denotes the *i*th vector of occurrences.

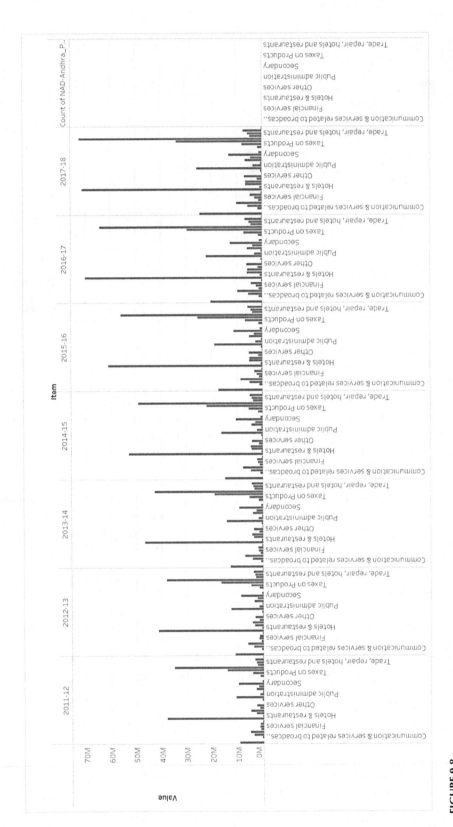

FIGURE 9.8
Crop statistics – Andhra Pradesh.

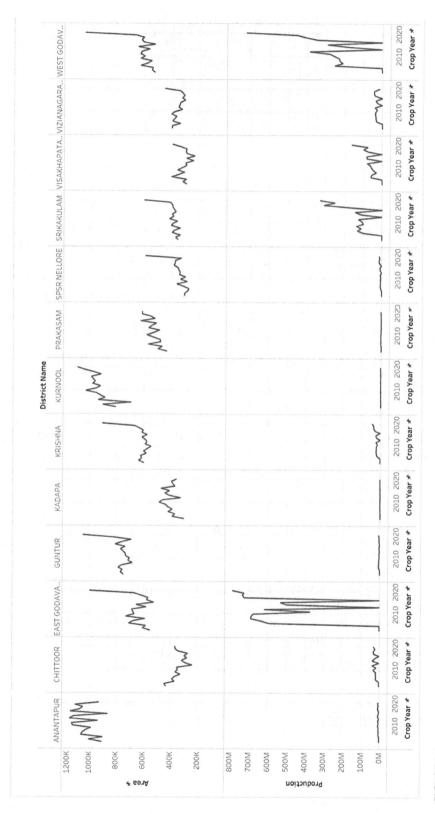

FIGURE 9.9
Crop yield district wise.

FIGURE 9.10
Sample crop maturity level.

Based on the analysis, the better crop yield shall be arrived using random forest technique and also regression analysis is more important since regression analysis is a method of predictive modelling approach that looks into the relationship between a dependant (target) and an independence (s) factor (predictor). For instance, if the crop is a seasonal crop, its maturity will be classified as Early-mid maturing, Mid-maturing, or Default. The harvesting period may also be forecast, allowing farmers to obtain a higher yield and higher profit. The sample is represented in the Figure 9.10.

The agricultural production and its annual ending monthly are illustrated in the below Figure 9.11.

The above graph Figure 9.12 represents annual crop output; utilising this, the farmer will be able to determine which product produces the best results, and accordingly, the farmers would seed and plant accordingly. The recommended and optimal crops for each zone are depicted in the diagram Figure 9.13.

The proposed technique aims at sustainable farming under different dimensions with a crop evaluation model. The crop evaluation numerical model has been created using four primary drivers: Market Value of the crop's goods, Service Expense for processing the goods to market, and Production cost to produce the crop. The relationship was built on the functional relationship between the variables. It is self-evident that these will be factors.

$$C_e = S_e + P_c + P_d + U_i + P_p \tag{9.3}$$

C_e – Crop evaluation
S_e – growth in crop per quantity of time lead to service costs associated with processing crop units to make unit goods
P_c – a rise in crop per time unit due to the expense of producing crop products
P_d – decrease in crop per time unit due to crops unit production
U_i – growth in crop every unit of time as a result of urbanisation
P_p – crop each unit of time growth due to population per unit time

Depending on the soil conditions and recommendations, the above Figure 9.14 depicts the predicted crop typically delivers a higher yield.

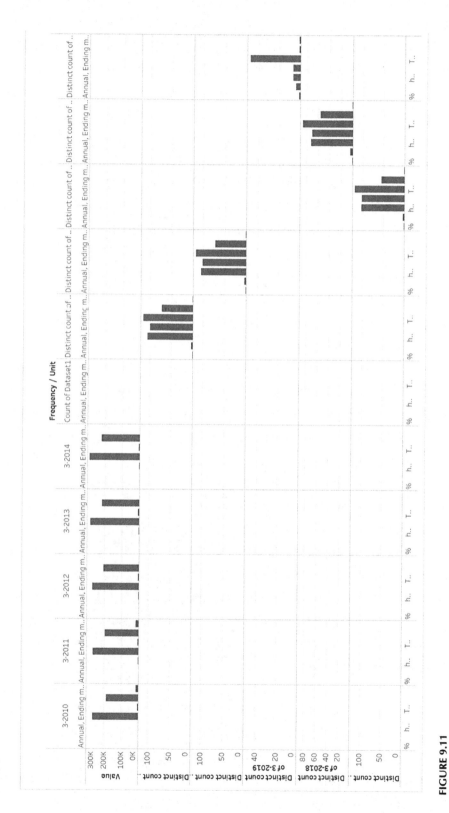

FIGURE 9.11
Agricultural production with annual ending.

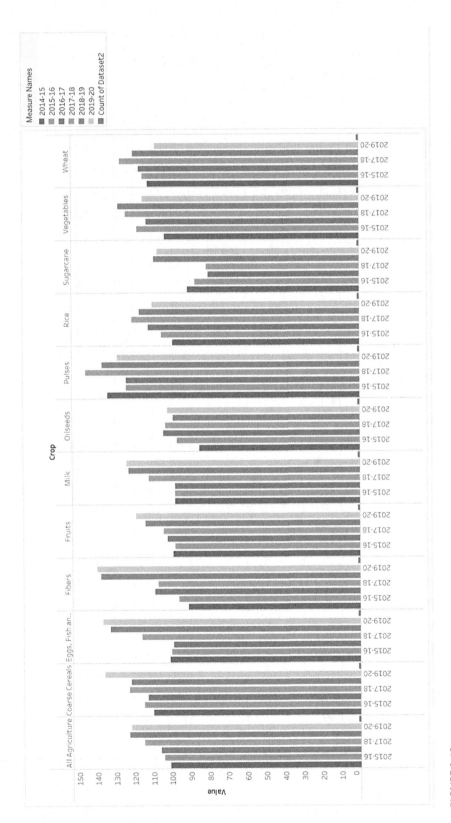

FIGURE 9.12
Crop production.

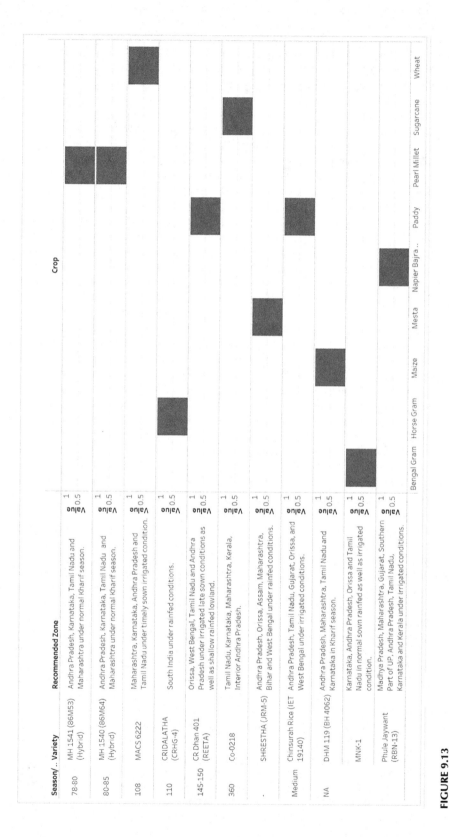

FIGURE 9.13
Recommended and optimal crops for each zone.

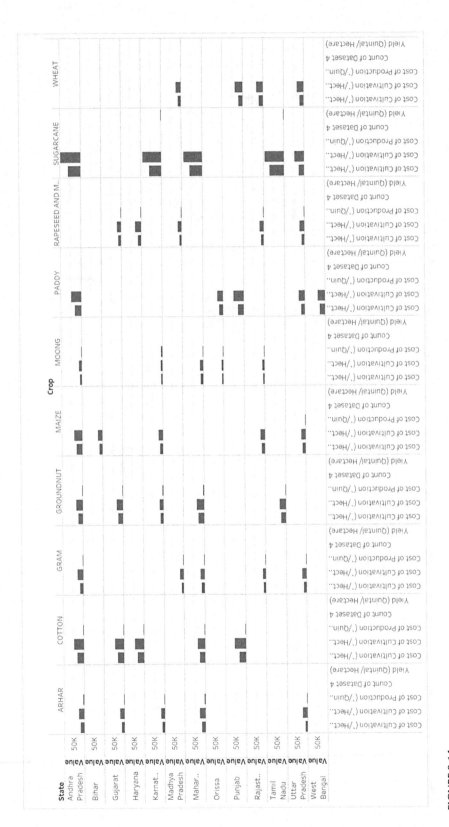

FIGURE 9.14
Crop forecasting for a greater yield.

9.5 Conclusion

Modern technology such as the IoT and big data must be used to better a domain like agriculture. It assists farm owners in establishing smart farming, which allows them to obtain more precise information on both in and out of the agriculture. Many of those can be rectified if the most reliable and realistic sensors were available, as well as internet access in all portions of the agricultural field. As a conclusion, we infer that the population of Indians and the globe is expanding exponentially, and the entire globe is demanding food. As an outcome, we must cultivate more agricultural crops to satisfy the world's demands by using sophisticated agricultural technologies.

The chapter focuses on the critical demands of modern agricultural development, and provides a smart farming technique to emulate the exact management and oversight of agricultural activities utilising IOT technologies and big data. The implemented framework makes use of IoT-enabled sensors, big data techniques such as regression and random forest algorithms, and a crop evaluation model for sustainable agriculture. The concept of greener, individuals, and environmental sustainability in a smart city has prompted global cities to think ahead in a smart way. In future, we plan to incorporate Artificial Intelligence (AI) to the proposed approach in order to establish a smart and automated application for effective farming, with AI algorithms driving wise strategic decisions.

References

1. Gandhi, V. P. and Zhou, Z. (2014). Food demand and the food security challenge with rapid economic growth in the emerging economies of India and China. *Food Research International*, 63, 108–124.
2. Kamilaris, A., Kartakoullis, A., and Prenafeta-Boldú, F. X. (2017). A review on the practice of big data analysis in agriculture. *Computers and Electronics in Agriculture*, 143, 23–37.
3. Saiz-Rubio, V. and Rovira-Más, F. (2020). From smart farming towards agriculture 5.0: A review on crop data management. *Agronomy*, 10(2), 207.
4. Ganesh Kumar, M., Thombare, V. J., Katariya, M. M., Veeresh, K., Raja, K. M. P., and Gopi, H. N. (2016). Non-classical helices with cis Carbon–Carbon double bonds in the backbone: Structural features of α, γ-hybrid peptide foldamers. *Angewandte Chemie*, 128(27), 7978–7982.
5. Khattab, A., Habib, S. E., Ismail, H., Zayan, S., Fahmy, Y., and Khairy, M. M. (2019). An IoT-based cognitive monitoring system for early plant disease forecast. *Computers and Electronics in Agriculture*, 166, 105028.
6. Alnafessah, A. and Casale, G. (2020). Artificial neural networks based techniques for anomaly detection in Apache Spark. *Cluster Computing*, 23(2), 1345–1360.
7. Elteir, M., Lin, H., and Feng, W. C. (2010, December). Enhancing mapreduce via asynchronous data processing. In *2010 IEEE 16th International Conference on Parallel and Distributed Systems* (pp. 397–405). IEEE.
8. von Cossel, M., Amarysti, C., Wilhelm, H., Priya, N., Winkler, B., and Hoerner, L. (2020). The replacement of maize (Zea mays L.) by cup plant (Silphium perfoliatum L.) as biogas substrate and its implications for the energy and material flows of a large biogas plant. *Biofuels, Bioproducts and Biorefining*, 14(2), 152–179.

9. Dhamodharavadhani, S. and Rathipriya, R. (2019). Region-wise rainfall prediction using mapreduce-based exponential smoothing techniques. In *Advances in Big Data and Cloud Computing* (pp. 229–239). Springer, Singapore.

10. Suryanarayana, C. (2001). Mechanical alloying and milling. *Progress in Materials Science*, 46(1–2), 1–184.

11. Ravikanth et al. (2021). An automated rescue and service system with route deviation using IoT and blockchain technologies. In *2021 IEEE Mysore Sub Section International Conference (MysuruCon)*. IEEE.

12. Kumar, T. S. et al. (2021). The improved effectual data processing in big data executing map reduce frame work. In *2021 IEEE Mysore Sub Section International Conference (MysuruCon)*. IEEE.

13. Ch, H. M. (2010). Changing scenario of organic farming in India: An overview. *International NGO Journal*, 5(2), 034–039.

14. Singh, P., Shahi, B., and Singh, K. M. Spread of Integrated Pest Management.

15. Ray, P. and Chowdhury, S. (2015). Kisan call centre: A new vista for Indian agricultural extension system. *International Journal of Social Sciences*, 4(2 and 3), 171–183.

16. Schönfeld, M., Heil, R., and Bittner, L. (2018). Big data on a farm—smart farming. In *Big Data in Context*. T. Hoeren, B. Kolany-Raiser, Eds., 109–120.

17. Han, J. and Kamber, M. (2012). *Data Mining: Concepts and Techniques*, Third Edition, Morgan Kaufmann Publishers, San Francisco.

18. Archana, S. and Elangovan, K. (2013). Survey of classification techniques in data mining. *International Journal of Computer Science and Mobile Applications*, 2(2), 65–71.

19. Shmueli, G., Patel, N. R., and Bruce, P. C. (2017). *Data mining for business analytics: concepts, techniques, and applications in R*, First Edition, John Wiley & Sons.

20. Dunham, M. H. (2006). *Data Mining Introductory and Advanced Topics*, First Edition, Pearson Education India.

21. Kotsiantis, B., Zaharakis, I. D. and Pintelas, P. E. (2007). Machine learning: A review of classification and combining techniques. *Artificial Intelligence Review*, 26(3), 159–190.

22. Raj, A., Bincy, G., and Mathu, T. (2012). Survey on common data mining classification techniques. *International Journal of Wisdom Based Computing*, 2(1), 12–15.

23. Sahu, H., Shrma, S., and Gondhalakar, S. (2011). A brief overview on data mining survey. *International Journal of Computer Technology and Electronics Engineering*, 1(3), 114–121.

24. Nabi, D. L. A. and Ahmed, S. S. (2013). Survey on classification algorithms for data mining: (Comparision and Evaluation). *Computer Engineering and Intelligent Systems*, 4(8), 18–24.

25. Barros, R. C., Basgalupp, M. P., Carvalho A. D., and Freitas, A. A. (2012). A survey of evolutionary algorithms for decision-tree induction systems, man, and cybernetics, Part C: Applications and reviews. *IEEE Transactions*, 42, 291–312.

26. Pernkopf, F. (2005). Bayesian network classifiers versus selective K-NN classifier. *Pattern Recognition*, 38, 1–10.

27. Maniya, H., Hasan, M., and Patel, K. P. (2011). Comparative study of naïve bayes classifier and KNN for tuberculosis. *Proceedings of the International Conference on Web Services Computing (ICWSC)*, 2(1), 22–26.

28. Mollazade, K., Omid, M., and Arefi, A. (2012). Comparing data mining classifiers for grading raisins based on visual features. *Computers and Electronics in Agriculture*, 84, 124–131.

29. Ritschard, G. (2013). CHAID and earlier supervised tree methods. In *Proceedings of the Contemporary Issues in Exploratory Data Mining in the Behavioral Sciences*, 70–96, Routledge.

30. Yuxun, L. and Niuniu, X. (2010). Improved ID3 algorithm. *Proceedings of the IEEE 3rd International Conference on Computer Science and Information Technology, 8*, 465–468.

31. Quinlan, J. R. (1986). Induction of decision trees. *Machine Learning*, 1(1), 81–106.

32. Quinlan, J. R. (2014). *C4. 5: Programs for Machine Learning*, Elsevier.

33. Steinberg, D. and Colla, P. (2009). CART: Classification and regression trees. In *The Top Ten Algorithms in Data Mining*, X. Wu, V. Kumar, eds. CRC Press, Taylor and Francis Group, 179–201.

34. Kumar, R. and Verma, R. (2012). Classification algorithms for data mining: A survey. *International Journal of Information and Engineering Technology*, 1, 24–48.

35. Carbonell, I. M. (2016). The ethics of big data in big agriculture. *Internet Policy Review*, 5(1), 1–13.
36. Diebold, F. (2012). Big Data, Pier Working Paper Archive, Penn Institute for Economic Research, 12–37.
37. Pham, X. and Stack, M. (2018). How data analytics is transforming agriculture. *Business Horizons*, 61(1), 125–133.
38. Xie, Jiong, Yin, S., Ruan, X., Ding, Z., Tian, Y., Majors, J., Manzanares, A., and Qin, X. (2010). Improving MapReduce performance through data placement in heterogeneous hadoop clusters. In *Proceedings of the IEEE International Symposium on Parallel & Distributed Processing, Workshops and Ph.d Forum (IPDPSW)*, 116, April 19 2010 to April 23 2010, Atlanta, GA.
39. Shim, K. (2013). MapReduce algorithms for big data analysis. In *Proceedings of the International Workshop on Databases in Networked Information Systems*, 44–48, Aizu-Wakamatsu Japan.
40. Yadav, N. Sudhakar, Srinivasa, K. G., and Reddy, B. Eswara. (2019). An iot-based framework for health monitoring systems: A case study approach. *International Journal of Fog Computing (IJFC)*, 2(1), 43–60.
41. Yadav, N. Sudhakar, Reddy, B. Eswara, and Srinivasa, K. G. (2018). An efficient sensor integrated model for hosting real-time data monitoring applications on cloud. *International Journal of Autonomic Computing*, 3(1), 18–33.
42. Yadav, N. Sudhakar, Rao, Mallikarjuna, Parameswari, D. V., Soujanya, K. L. S., and Latha, Challa Madhavi (2021). Accessing cloud services using token based framework for IoT devices. *Webology*, 18(2).
43. Dinkins, C. and Jones, C. (2019). Interpretation of soil test reports for agriculture. *MT200702AG, Montana State University Extension: Bozeman, MT, USA*, 2013.
44. Santhi, P. V., Kapileswar, N., Chenchela, V. K., and Prasad, C. V. S. (2017). Sensor and vision based autonomous agribot for sowing seeds. In *2017 International Conference on Energy, Communication, Data Analytics and Soft Computing (ICECDS)*, pp. 242–245. IEEE.
45. Zhang, L., Dabipi, I. K., Brown Jr, W. L. (2018). Internet of things applications for agriculture. *IoT A to Z: Technologies and Applications*, 507–528.

Index

A

Aarogya Setu, 160
Ad Hoc Networks (VANETs), 122
Advanced Vehicle Control Systems
    (AVCS's), 133
Akaike Information Criterion (AIC), 116
artificial neural network (ANN), 113, 180

B

Back-propagation Neural Network (BPNN), 182
Backpropagation-neural network (BP-NN), 114
Big data, 3, 69, 78, 160
    Architecture, 7
    Characteristics, 6
Big Data-IoT, 104, 183
Blockchain, 109

C

C4.5, 179
Cipher Text Policy-Attribute-based Encryption
    (CP-ABE), 160
Cloud computing, 24
Cloud-IoT, 33, 34
Congestion-aware routing algorithm
    (CARA), 138
Constrained Application Protocol (CoAP), 9
COVID-19, 159

D

Deep Neural Network (DNN), 19
Dynamic Throughput Maximization System
    (D-TMF), 126
Dynamic traffic light sequencing (DTLS), 92
Dynamic Traffic Management Centre
    (DTMC), 138

E

Emergency vehicles (EV), 121
Enhanced Intelligent Driver Module
    (EIDM), 137

Enhanced Sampling Rate (ESR), 133
Ensemble learning, 111

F

Floating car data (FCD), 124
Fog computing, 11
Forensics-as-a-Service (FaaS), 4
Fuzzy C Means (FCM) algorithm, 122
Fuzzy K Means (FKM) algorithm, 122

G

Gateways, 86
Green Corridor, 138

H

Hidden Markov Model (HMM), 167
Hypertext Transfer Protocol (HTTP), 167

I

IBM Blue-Mix, 132
Industrial IoT (IIoT), 11
Information and Communication Technology
    (ICT), 2, 108, 179
Intelligent traffic congestion control
    (ITCC), 138
Intelligent Traffic Management Systems, 78, 81
    Architecture, 85
Intelligent Transport Systems (ITS), 78
Internet bio-nanothings (IoBNT), 166
Internet of Everything (IoE), 37
Internet of Things (IoMT), 163
Internet of Things (IoT), 2, 8, 26, 43, 58, 76
Internet of Vehicles (IoV), 11
IoT challenges, 47

L

Long Short-Term Memory (LSTM), 181
Long Short Term Memory Models (LSTMs), 125
Low-Power Wide-Area Network (LPWAN), 8

M

Machine-to-Machine Communication (M2M), 34
Malfunction Indicator Light (MIL), 130, 133
Mean Absolute Percentage Error (MAPE), 181
Message Queuing Telemetry Transport
    (MQTT), 167
Microcontroller Unit (MCU), 124
Mobile multimedia agricultural advisory
    system (MMAAS), 180
Multi-step authentication, 51

N

Narrow Band-Internet of Things (NB-IoT), 9

O

OBD2 (On Board Diagnostic-2), 130

P

Patient Health Portals (PHP), 128
Predictive analytics, 30

Q

Quality of Life (QoL), 2

R

Radio frequency identification (RFID), 8, 78,
    110, 139
Random forest, 186
Reinforcement Learning, 114
Remote Patient Monitoring (RPM), 165
Role-based Access Control Model (RBAC), 160
Root Mean Square Error (RMSE), 181
RT PCR, 164

S

Semi-Supervised Learning, 113
Slave road side units (SRSU), 138
Smart city, 2, 9, 58
    Architecture, 61, 71, 140
    Challenges, 94
Smart farming, 184
    Sensors, 184
Smart intelligent transportation system, 112
Smart Irrigation, 185
Smart teaching learning, 68
Smart waste management, 67
Smartphones, 80
Social IoT (SIoT), 125
Software-Defined Network (SDN), 63
Supervised Learning Algorithm, 112
Support Vector Machines (SVM), 113
Sustainable agriculture, 188

T

Traffic Management Controller (TMC), 86
Traffic Monitoring Unit (TMU), 86

U

Unmanned aerial vehicle (UAV), 168
Unsupervised Learning, 113

V

Village Level Entrepreneur (VLE), 178

W

Web ontology language (OWL), 62
Wireless Body Area Networks (WBANs), 128
Wireless Sensor Networks (WSN), 136

Printed in the United States
by Baker & Taylor Publisher Services